# 作者自述(About the author readme)

(摄于美国辛辛那堤)
(Cincinnati,USA)

1996 年退休,高级工程师。1996—1984 年,任地矿部石油地质中心实验室测试主任工程师;1979—1977 年,任地矿部湘中石油地质攻关会战地化组长;1976—1974 年,编撰《石油地质实验基础》;1974—1961 年,任地矿部三普实验室副主任、主任、技术负责,其间任长庆油田会战吴战一井副指挥;1957 年参加国家统一高考,考入西北农学院林学系,后因政治原因辍学,靠校外自修;1953 年毕业于南京地质学校,1954 年修完俄语速成班。

从 1959 年起,由矿物岩石化学分析转行石油地质实验,开始从事石油地球化学和石油成因研究。创立的石油成因新理论——湿封闭体系微生物和碳酸盐主导生油理论,得益于长期深入石油地质野外实践和油田实践,从普查到勘探再到开发,丰富了宏观石油地质经验。与此同时,担任实验室技术负责,历时 30 多年,审查全国各地区石油地质各类样品,包括中外合作交流与专题研究的实验分析测试报告达数十万份,各类分析数据达百万个,具有极其特殊的宏观与微观紧密结合、相得益彰的研究能力与经历经验。创立这种石油成因新理论是长期大量科技创新的必然结果。

非常感谢内弟宏昌、弟媳惠君资助本书出版。

作者相关讯息:

E-mail:dasenheabcde@ 163. com

腾讯微博:何志高

# 湿封闭体系微生物和碳酸盐主导生油理论
# 论文选集

# The Theory of Oil-Generation Driven by Microorganism and Carbonate in the Waterlogged and Closed System

**Selected Papers**

**编著者** 何志高
Author He Zhigao

苏州大学出版社
Soochow University Press

**图书在版编目(CIP)数据**

湿封闭体系微生物和碳酸盐主导生油理论论文选集：汉、英 / 何志高编著. —苏州：苏州大学出版社，2014.6
ISBN 978-7-5672-0884-1

Ⅰ.①湿… Ⅱ.①何… Ⅲ.①石油成因—文集—汉、英 Ⅳ.①P618.130.1—53

中国版本图书馆 CIP 数据核字(2014)第 109054 号

**湿封闭体系微生物和碳酸盐主导生油理论论文选集**
何志高 编著
责任编辑 倪 青

苏州大学出版社出版发行
(地址：苏州市十梓街 1 号 邮编：215006)
苏州恒久印务有限公司印装
(地址：苏州市友新路 28 号东侧 邮编：215128)

开本 700 mm×1 000 mm 1/16 印张 13.75 字数 247 千
2014 年 6 月第 1 版 2014 年 6 月第 1 次印刷
ISBN 978-7-5672-0884-1 定价：68.00 元

## 内容简介

石油成因是世界性的科学难题。创立科学的石油成因理论，不仅是世界科学界的重大事件，而且会极大地提高石油勘探开发效率。本书系统地汇集了作者40多年来研究创立的石油成因新理论——湿封闭体系微生物和碳酸盐主导生油理论的相关论文20篇，具有推动世界地质科学发展的重大意义，对当今全世界热衷于页岩气的研究、开发具有参考和指导作用；而且，非常有趣地发现，该理论可能对探索地球外天体有无地球类人类提供新的研究判断依据。

石油究竟是怎么生成的？这个世界争论了200年的科学难题，本书作者做出了随时间发展而不断深入研究至走向成熟的论述。

## About the book

The origin of petroleum is a worldwide scientific problem. Creation of a new scientific theory of petroleum genesis, is not only a major event in the world of science, and will greatly improve the efficiency of oil exploration and development. This book brings together a systematic study of the creation of life, 40 years of oil causes a new theory "The Theory of Oil-Generation Driven by Microorganism and Carbonate in the Waterlogged and Closed System", the relevant 20 papers, has great significance to promote the scientific development of the world's petroleum geology; Keen on the world today shale gas, also as a reference and guidance; also, very interesting to find that the theory might explore whether the Earth's Celestial Class extraterrestrial humans, according to a new research judgment.

The author of this book to tell you!

How to origin, Petroleum?

突破
祷告中的缺乏
余孝礼/著

# 前言(Foreword)

关于湿封闭体系微生物和碳酸盐主导生油论，如果从1986年笔者在《地质学报》上发表首篇论文算起，至今进入成熟阶段，共用了28年。

顾名思义，该理论确定，石油的生成，包括油田的形成，下述4个元素、4项条件或称4位并列体是必须同时具备、缺一不可的：

1. 湿的、饱含水的体系；
2. 封闭的、能形成还原环境的体系；
3. 微生物的主导作用，微生物对沉积有机生物物质的分解、改造和直接生烃作用；
4. 碳酸盐的匹配协同主导作用。

理论名称的变化更有利于反映理论本质。

历史上的经验教训是极为深刻的。1940年至1952年，美国石油地质研究所研究课题43号方案对石油成因和原油的组成进行了长达13年的研究，投入巨大的人力和资金。随后出版的《石油烃化学》一书中所列出的汽油馏分单体烃类定性定量分析结果令人极为感叹其科学技术水平之高。8个碳原子的全部同分异构体的分析精度(体积百分比)达到小数点后面第2位。然而，43号方案的结论却是“沉积物中有机物质的微生物分解和放射性轰击没有对石油的生成起到重要作用”。这个结论对早期成油论是一个致命性打击。由此也就在世界范围掀起了石油晚期成油学说热，并且该理论的统治时间达60年之久。但是，前不久，美国石油地质学家J. M. Peters等应用$^{14}C$测年法确定加利福尼亚海湾Guaymas盆地的石油和碳酸盐的生成年龄为4240~5705年。这是非常有意义的重大发现。这项成果既是早期成油论的有力证据，当然也是对湿封闭体系微生物和碳酸盐主导生油论的强有力支持。因此，笔者也深信，新理论一定会得到美国等国际石油界的关注和肯定，就像笔者以前的文章发表后美英等国学者来函索取论文一样。

笔者认为，本书论文跨年度长达48年(1965—2013)，历史悠久，篇篇都是创新之作，信息量大，还有大量的未编入书中的文章、专题报告以及执笔的集体之作，都有新理论的创作源泉。但是，要比较清楚地认识该理论，本书中的以下4篇文章是必须详细弄清楚的：

1.《论石油烃类气体成因》：这是一篇当年中国科学院资深院士程裕淇做

出很高评价之作，他亲自帮助修改英文稿。该文十分肯定石油气态烃类是微生物直接脱羧形成的，从而揭开新理论创作之序。

2.《湿封闭体系综合效应成油论》：这是一篇在国际会议上宣讲的大会发言论文，是整个地矿部唯一的一篇。会后，该文被 *Journal of Southeast Asian Earth Sciences* 收录刊登，主编对该文的评价是：非常有趣而新颖。随后，美英学者来函索取论文。该文奠定了湿封闭体系和微生物重要作用的坚实基础。

3.《中国"加瓦尔胚胎"与有机成油理论铁证生油剖面　一种新命名的岩石——"油泥灰岩"命名申述》：这篇文章被《石油知识》编辑部邓主任称赞为好文章，以头版头篇、特大字体标题、关注和编者按 4 项高规格评价刊出。这篇文章把湿封闭体系成油理论从原来的以实验室为主、微观为主上升到了有一条地质体中厚度近千米全岩心宏观与微观兼备的综合剖面；中国"加瓦尔胚胎"概念的提出与对比阐述，其意义也是非凡的；还有"油泥灰岩"的命名申述，都是创新之作。

4.《论石油形成体系的化学基础——化学反应方程集合——油田形成的最大奥秘》：这是最近的可能也是最后的大块文章，从题目就可以看出文章的高度。这篇文章全部概括了前述的石油形成的 4 项必备条件的理论基础，更加突出了碳酸盐的作用。应该看到，碳酸盐除了能消除生成烃类的化学反应同时生成的二氧化碳，确保该生烃反应能一直进行下去之外，还可为沉积物沉积之前的水体生物（水生植物和水生动物）繁衍提供各种物质来源。可以这样说，一个盆地内的巨大油田一定在沉积建造相应的原生储油层时，有一个非常繁盛的水生生物水体世界。因为碳酸盐古物源，为水体提供了光合作用所需的二氧化碳，提供了新生态氧，提供了高效生化反应生物代谢的纳米级碳酸钙、碳酸镁以及生物生命所必需的微量元素和化合物等。

笔者把碳酸盐对湿封闭体系所做的贡献称为"微生物主导作用的协同匹配主导作用"；而碳酸盐对沉积水体生物以及生油原始有机物质的数量的贡献是另一个极具重要意义的石油形成包括油田形成最大奥秘的组成部分。当然，还有像世界上最大油田加瓦尔油田那样的碳酸盐本身还是一种碎屑沉积物，还构建特殊的像卡斯特那样的湿封闭体系，可以想象，那种碳酸盐及其沉积环境必然是一种特殊的生油最佳条件。

很遗憾的是，本书出版时已是笔者力不从心之时，很多错误未能发现并改正，尤其是英文。补救的办法是寻到原文核对。但是，这是真古迹，是"半个世纪、一种理论、一代人生"的真迹。

何志高

2014 年 5 月

# 目 录 Contents

# 1. 湿封闭体系综合效应成油论

## Wet closed system to generate petroleum theory

何志高

(地质矿产部石油地质中心实验室,江苏无锡,214151)

**摘要:** 本文提出的湿封闭体系综合效应成油论,主要强调饱含水的封闭的特定地质体系综合效应成油、微生物的直接成油及压力作用的重要性。微生物直接成油作用不是发生在甲烷的生成阶段,而是在此阶段的后期,是一种特定的生物化学反应。这是区别于其他生物成油理论的一个标志。

本理论是在研究了原油、高压地层原油(包括油溶气)、沼气及其发酵溶液与发酵过程(包括某些发酵机制)、碳酸盐岩酸解气、油田气、近代与古代沉积物等多种资料,并进行人工沼气和生物作用模拟实验的基础上提出来的。

本石油成因理论将成油的过程分成两个阶段,即成油前期的准备阶段与成油阶段。本文主要从十个方面进行论述。

### 1 基本论据

#### 1.1 细粒碎屑沉积物中有机物质的重新分配

图1是仿生物作用、物理分异作用的模拟实验的素描图。图1表明了有机物质随同各种碎屑物沉积后,在有水存在的条件下,受生物作用以及物理分异作用重新分配。轻的、颗粒小的有机碎屑在上部;最轻的、细分散状的及胶体状的有机物聚集于沉积物的表面(箭头所指处)。

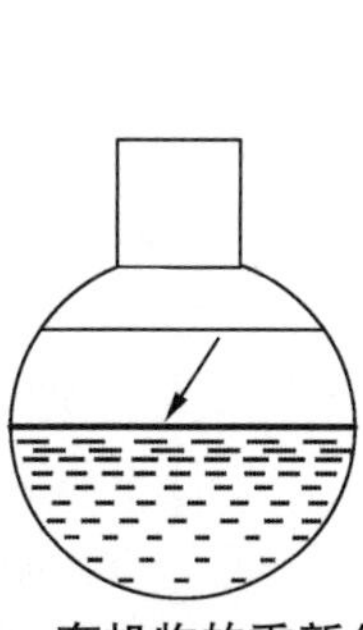

图1 有机物的重新分异作用的模拟实验

#### 1.2 准备阶段的生化反应无石油烃类生成

处于水介质条件下的细分散状的有机物质,无疑进行着生物化学作用。这种作用已被许多研究资料证实。如 Brassell 等指出(1984),尤其在水层和地表沉积中,微生物的作用是主要的,它们在甾族化合物的异构化、结构重排、芳构化方面起了重要作用。根据沼气的发酵研究可以看到,在生物化学的初始阶段,主要生成甲烷和脂肪酸,无石油烃类生成。

表 1 是人工沼气、石油气甲烷碳同位素及气体组成资料对比。从表 1 可见,随着发酵时间的延长,$\delta^{13}C$‰值增大,二氧化碳碳同位素的变化比甲烷的变化更为明显。发酵 19 d,甲烷尚未变到石油类甲烷的碳同位素值范围。二氧化碳碳同位素值、$CO_2/CH_4$ 值亦如此。发酵产气中亦未见到乙烷以上重烃的显著含量。因此,可以认为,在这个准备阶段,没有石油烃类生成。

**表 1　沼气、天然气分析资料对比**

| 发酵时间 (d) | $CH_4$ $\delta^{13}C$‰ (PDB) | $CO_2$ $\delta^{13}C$‰ (PDB) | $\frac{CO_2}{CH_4}$ (体积比) | $\frac{C_2^+}{\sum HC}\cdot 100$ (%) | $CO_2$ $\delta^{13}O$% (PDB) |
|---|---|---|---|---|---|
| 2 | -53.48 | -13.47 | 0.56 | <0.001 | -4.88 |
| 5 | -54.69 | -10.45 | 0.70 | <0.001 | -4.85 |
| 7 | -53.95 | -8.51 | 0.87 | <0.001 | -4.79 |
| 9 | -53.33 | -7.85 | 1.14 | <0.001 | -4.63 |
| 13 | -52.92 | -7.50 | 1.19 | <0.001 | -4.69 |
| 15 | -52.75 | -8.20 | 1.22 | <0.001 | -4.88 |
| 17 | -52.13 | -8.12 | 1.17 | <0.001 | -4.39 |
| 19 | -48.01 | -6.77 | 0.93 | <0.001 | -4.51 |
| 江苏天然气 | -30.50 | -5.14 | 0.01* | 32* | -3.28 |

*油溶气资料。

### 1.3　石油烃类气起源于脂肪酸生物脱羧

脂肪酸系生物化学反应的必然产物之一,不仅为很多研究者所定论,而且沼气发酵液的分析早已予以肯定。在饱含水的封闭体系中,从甲酸至丁酸($C_1$—$C_4$)可以与水完全互溶,$C_5$—$C_7$ 在水中有较大的溶解度。由于封闭体系中饱含水分,$C_8$—$C_{10}$ 等酸类在水中的溶解量不可低估。这些脂肪酸在封闭体系中发展变化的地球化学前景可以有以下几种情况:

(1) 热分解的可能性小

根据脂肪酸的化学性质,低级脂肪酸在常压下蒸馏不分解,即在沸点温度时保持稳定。乙酸沸点 118 ℃,丁酸 163 ℃。由此可见,在通常石油存在的温度下,不会有脂肪酸的热分解、热脱羧作用发生。原油、油田水、古代深部岩石中均已检测出脂肪酸，这是一种有力的佐证。

此外，从大量的地球化学研究资料可见，岩石中的脂肪酸的奇偶碳数化合物的分布与烃类化合物奇偶碳数分布，随埋深增加其变化趋势不匹配。

假定岩石中的烃类主要是由脂肪酸热脱羧形成的，那么，必定是烃类的奇碳优势随埋深增加而增大(或是偶碳优势降低)。事实上，在随深度增加的地质剖面，烃类是奇碳数降低或偶碳数增加(膏盐环境等特殊情况另论)，并向奇偶比值接近于 1 的方向变化。因此，该假定与事实不符，不能成立。烃类不是由脂肪酸热脱羧形成的。

(2) 生物降解的可能性较小

沼气发酵中普遍存在着脂肪酸的生物降解(即大分子脂肪酸规则地降解为小分子脂肪酸)作用，而且人们期待增强这种降解作用，以达到增加甲烷产率的效益。

在有水的封闭体系中，虽然不能否定生物降解作用，但是，由于增加了氢气分压(氢气分压增加是必然的)，降解反应的生成自由能变化值($\Delta G$)正值增大，降解作用难以进行(见图 2)。并且，这种氢气效应随脂肪酸的碳数增加而增大。因此，可以认为，生物降解反应不是脂肪酸变化的主要反应形式。事实上，从石油富含重烃类的组成特点来看，也不能认为有此种显著的作用发生。

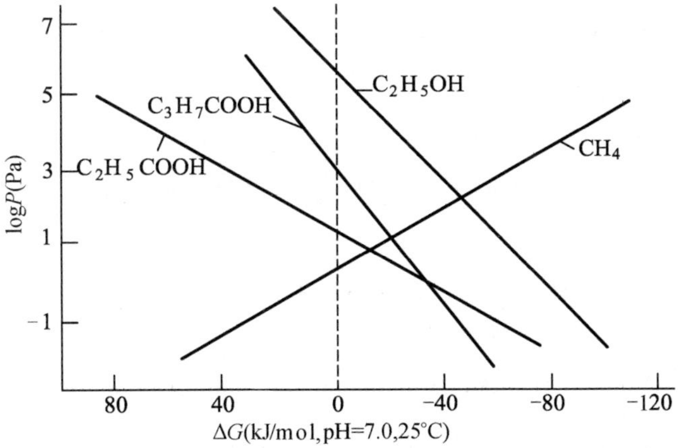


**图 2　氢分压($P$)对乙醇、丙醇、丁醇的降解和形成甲烷的自由能变化($\Delta G$)影响(浓度 $10^{-13}$ mol)**

(3) 生物脱羧生成烃类

笔者发现，沼气发酵液中的脂肪酸按特定的异构体规则处理，脱羧后形成的烃类的组成与石油气(选取能反映石油气原始组成特性的高压地层原油中油溶气态烃为代表)组成之间存在着高度的对应相关性。这种相关性主要表现在石油气与脂肪酸两者含量变化曲线的形态一致(图 3)以及两组数据的很大的线性相关系数,其值为 0.9935。

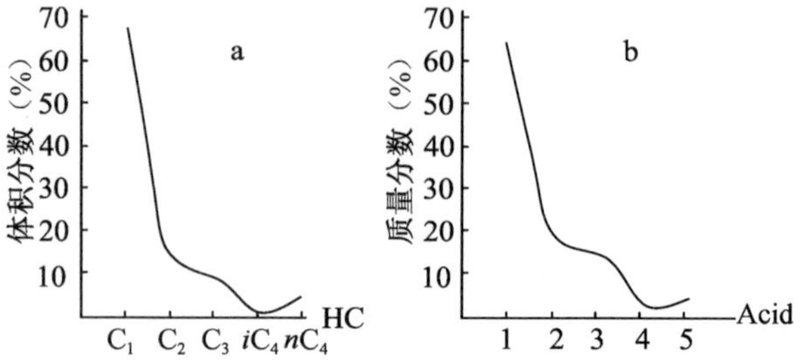


**图 3　油溶气与脂肪酸组成分布曲线**

a. 油溶气　b. 脂肪酸　b 图中:1. $CH_3COOH$　2. $CH_3 \cdot CH_2 \cdot COOH$

3. {iso-butyric acid / *n*-butyric acid}　4. 1/2 iso-pentenoic acid　5. {*n*-pentenoic acid / 1/2 iso-pentenoic acid}

脂肪酸脱羧成烃的异构体规则,不仅在理论上可以做出解释,而且其本身对应相关性就是一种检验,其脱羧规则如下:

① 正丁酸(异丁酸)⟶丙烷

$$CH_3—CH_2—CH_2—COOH \longrightarrow C_3H_8 + CO_2$$

$$CH_3—\underset{\displaystyle CH_3}{\underset{|}{CH}}—COOH \longrightarrow C_3H_8 + CO_2$$

② 2 -甲基丁酸(正戊酸)⟶正丁烷

$$CH_3—CH_2—\underset{\displaystyle CH_3}{\underset{|}{CH}}—COOH \longrightarrow nC_4H_{10} + CO_2$$

$$CH_3—CH_2—CH_2—CH_2—COOH \longrightarrow nC_4H_{10} + CO_2$$

③ 3 -甲基丁酸⟶异丁烷

$$CH_3—\underset{\displaystyle CH_3}{\underset{|}{CH}}——CH_2—COOH \longrightarrow CH_3—\underset{\displaystyle CH_3}{\underset{|}{CH}}—CH_3 + CO_2$$

脂肪酸脱羧生成烃类的反应,以乙酸为例,反应的热力学性质可以做如表 2 所示的分析。

由表 2 可见,乙酸脱羧反应在热力学标准状况下是放热、熵增、自由能降低的反应。在含水的封闭体系中,温度与压力对上述各热力学状态函数的影响和修正值均较小,不可能导致反应倾向性的变化。

**表 2 乙酸脱羧反应的热力学分析**

| 参数 | $CH_3$—COOH(l)⟶ | $CH_4$(g) + | $CO_2$(g) | 参数变化值 |
|---|---|---|---|---|
| $\Delta Hf^0$ | -487.3 | -74.9 | -393.8 | $\Delta H = -18.7\ kJ \cdot mol^{-1}$ |
| $S^0$ | 159.9 | 186.3 | 213.8 | $\Delta S = 240.1\ J \cdot deg^{-1} \cdot mol^{-1}$ |
| $\Delta G_f^0$ | -392.7 | -50.87 | -394.7 | $\Delta G = -52.8\ kJ \cdot mol^{-1}$ |

根据 CRC 热力学规则及有关的资料，求得的丙酸及其更高分子脂肪酸脱羧反应，其吉氏自由能变化值 $\Delta G$ 均约为 -41 kJ/mol（对整体反应来说是 kJ/反应），它们均在生物生理变化能量级范围（32 ~ 50 kJ/mol）。

### 1.4 成岩阶段成烃

碳酸盐岩研究表明（参考文献 2 及何志高“碳酸盐岩找油的一个重要领域”1985），用酸分解碳酸盐矿物后所释放出来的烃类是一种封闭的、被矿物裹挟的烃类，是在碳酸盐矿物结晶成岩过程中生成的原生烃类。这些烃类如表 3 所示，其组成、异构体比值以及碳同位素比值均系石油烃类。由此可见，形成石油气态烃类的时间相当于碳酸盐岩的固结成岩期。

**表 3 碳酸盐岩酸解气与油田气对比**

| 项目 / 样品类 | $CH_4$ $\delta^{13}C‰$ (PDB) | $C_2H_6$ $\delta^{13}C‰$ (PDB) | $C_3H_8$ $\delta^{13}C‰$ (PDB) | $C_4H_{10}$ $\delta^{13}C‰$ (PDB) | $\frac{nC_4H_{10}}{iC_4H_{10}}$ (体积比) | $\frac{C_2^+}{\Sigma HC} \times 100$ (%) |
|---|---|---|---|---|---|---|
| 酸解气(T 岩心) | -53.05 | -28.17 | -26.57 | -26.09 | 1.13* | 22.1 |
| 酸解气(P 岩心) | -34.66 | -26.44 | -24.35 | | | |
| 江苏油田气 Ed | -46.28 | -32.80 | -29.41 | -28.75 | 2.8 | 32.2 |
| 江苏天然气 Ef | -37.93 | -27.81 | -26.22 | -26.43 | 2.5 | 9.9 |

* 24 个样品平均值。

### 1.5 油、气同生于同一封闭体系

油（气）勘探与开发的事实表明，多数工业油藏，有油必有气，油与气共存。本文所研究的苏北下第三系油田原油与油溶气具有同生的特点，油、气之间的关系极为密切，它们的生成条件一致，并且聚集、保存在一个储油构造之中。

这些结论是在系统地研究了苏北原油油源及高压地层原油与油溶气态烃的关系（何志高，苏北下第三系原油油源探讨，1981）之后得出的，主要论据有：

（1）油溶气的组成变化与油层压力的关系极大，而与油层温度、深度无直接的关系；

(2) 原油产层深度(温度)与原油正烷烃色谱主峰碳数毫无关系，与<250 ℃轻馏分量亦无关，其深度与馏分量之间的线性相关系数为-0.058。此种情况与 Yolande Califert 法(1968)的模拟实验资料一致，即原油在高温、高压条件下(230 ℃、$300\times10^5$ Pa、72 h)，没有 $C_1$—$C_5$ 的低分子烃类产生；

(3) 根据油溶气中二氧化碳、氮气的含量很低及其含量变化等特点和原油本身正构烷烃的分布特征，可以判断出所研究的原油未遭受生物降解和游离氧的氧化作用。

上述三种论据表明了溶解于原油中的气体的原生性。既然沸点、临界温度、临界压力均很低的原生的气体都能得以保存，并不扩散损失，那么，这个形成油气的体系必然具有封闭的性质。因此，石油生成必定需要一种封闭的体系。

### 1.6 压力效应

从大量的石油地质现象以及高压地层原油性质与地面原油性质对比，可以看到压力在石油运移、聚集等方面的重要作用。

图4的样品是鄂尔多斯盆地南缘四郎庙构造钻井深约800 m处取得的岩心(系硅质胶结物为主的粉砂岩，储油物性很差，有效孔隙率约10%，绝对渗透率<0.1 mD)。其中左图是新鲜岩心劈开面，裂隙中充填原油，裂隙周围无任何石油显示；右图是散落在井场附近土壤中达数年之久的与新鲜岩心相似的岩心的劈开面，裂隙周围以及整个岩石孔隙中充填着石油。这些现象相当普遍。四郎庙构造三叠系砂岩，发育着垂直层理与平行层理的裂隙，裂隙中含油，宽度约为数毫米。除此之外，碳酸盐岩亦具有类似现象。例如，鄂尔多斯盆地安定组泥灰岩、苏南奥陶系灰岩均能看到岩心放置一段时间后石油向裂隙周围扩散的情况。

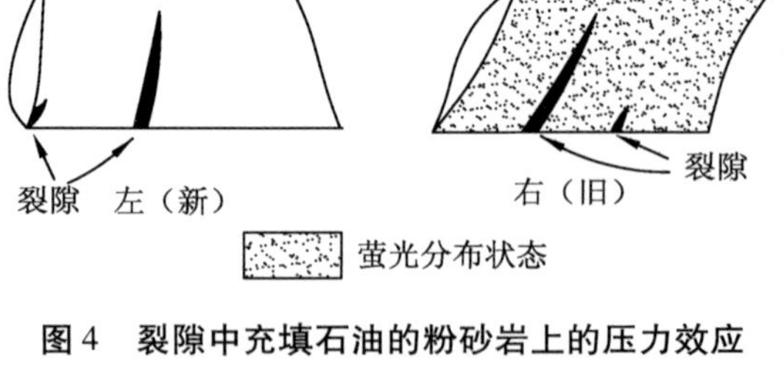


**图4 裂隙中充填石油的粉砂岩上的压力效应**

笔者认为，上述现象可以表明压力的作用导致石油在岩心中分布状态的变化。当压力高时，分散的石油聚集为聚集状态的石油，活动于岩石的裂隙通道之中，并且由于石油分子间增强的结合力，更强有力地结合在一起，因而很难向渗透率很低的裂隙围岩渗透；当压力降低以后，石油扩散渗透到物性很差的围岩。前者就是分散石油的聚集效应，后者就是聚集状态石油的分散。

图5是原油黏度与压力关系曲线。众所周知，黏度是分子间的内摩擦力指标，它的大小表示分子间作用力的强弱。不难理解，随着压力的增大，分子间距离缩短，其作用力增强，表征内摩擦力大小的黏度必然增高。对于富含非烃的原油来说，极性分子可产生的诱导偶极，导致分子间的作用力比高演化程度的原油更大。因此，压力尤其对初生态的分散的重油的聚集效应更大。

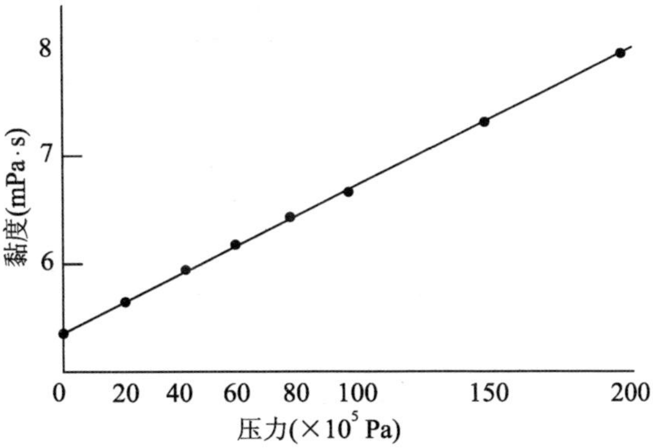


**图5 压力-黏度变化关系图**

另一方面，压力增加，低分子烃类液化程度与数量都随之增加，这对石油的运移、实现大规模的石油聚集是极有益处的。

### 1.7 热效应

随着沉积埋深作用的增强，温度的增高，热效应从小到大。这种热效应主要有以下三种形式：

（1）热降解作用增加烃类数量的效应

很多事实早已表明，无论原油本身或者岩石抽提物，随深度的增加，烃类数量增加。无疑，在封闭体系中，这种热降解作用是普遍存在的。

（2）增压效应

由于温度增高，烃类气化增加，气体膨胀，导致压力增大。

（3）降低石油黏度的效应

随温度的增高，石油黏度降低，无疑这种效应是很有利于石油运移的。

上述三种效应是显而易见的，无须赘述。

### 1.8 初生的、含水的封闭体系实例

通过对我国海洋一口井的现代沉积物（淤泥）覆盖下的第四纪粉砂岩的小间距取样的研究，业已发现，在淤泥底部的泥质粉砂岩是一个已有石油生

成的初生的含水封闭体系，这种石油处于0～1 m深处，如表4所示。

表4中资料表明，现代沉积物覆盖下的泥质粉砂岩，0～1 m深处的抽取物确系石油；随着粉砂岩往下部方向变化，石油类物质相应减少，这种变化可以看成是石油早期的向上运移的作用。由此可见，石油的生成时间是在成岩作用阶段；石油生成的场所是含水的封闭体系。同时还可看到石油早期的运移以及与石油生成紧密相联系的石油运移、聚集作用。

**表4　封闭体系成油典型实例资料**

| 样品位置 | 岩性 | 抽提物族组分 | | | | 芳烃红外光谱 $A_{810}/A_{750}$ |
|---|---|---|---|---|---|---|
| | | 饱和烃（%） | 芳烃（%） | 胶质（%） | 沥青质（%） | |
| 0～1 m | 泥质粉砂岩 | 52.0 | 8.8 | 37.5 | 3.5 | 0.82 |
| 1～2 m | 泥质粉砂岩 | 24.0 | 5.5 | 62.5 | 7.9 | 0.70 |
| 4～5 m | 泥质粉砂岩 | 9.7 | 4.9 | 75.7 | 9.8 | 0.50 |

### 1.9　油田实际

运用饱含水的封闭体系综合效应成油理论，可以阐述大量的石油地质问题。现略举例如下：

（1）很多“不成熟”原油以及找不到对应的“生油岩”的油田均可获得合理的解释。这是不言而喻的，因为本理论认为，原油从生成后，随即就有运移、聚集作用，且这些石油是富含胶质与沥青质的，故而显示出“不成熟”性。本理论认为，石油不是从“生油岩”中生成后向储集岩中运移的，当然不会存在“富含有机质的致密的生油岩”。

（2）据油田中的大量碎屑岩样品的地球化学分析结果表明，凡是砂质或含砂质的泥岩，生油指标较高。我国泌阳油田的资料尤有说服力。这些规律的起因可以归纳为：砂质或含砂的沉积物比泥质沉积物具有较大的孔隙度，因而含水较多，微生物对沉积有机物质的改造作用时间更长，改造的程度更深，生油作用更彻底，生油指标也就更高。

（3）碳酸盐岩地区沉积相变带富含石油的规律①。碳酸盐岩地区介于碎屑与碳酸盐沉积之间的相变带，由于具有较高的孔隙度，因而含水较多，同理，生成石油的作用更强烈。因此这种体系必然富含石油。

总之，凡是干酪根热降解成油理论难以解释的问题，诸如苏北油田的油源问题、冀中一些油田的石油储量等问题都可以得到合理的解释。

---

①　何志高，窄而厚的相变带沉积物富含石油，1985。

### 1.10 油藏形成的伴生迹象

油藏形成时，它必定伴随着岩性、水及周围介质的特定的物理及化学变化。能正确解释这些变化是对油气成因理论的一种检验。

本文主要解释下列一些伴生迹象：

（1）油田水中脂肪酸含量较高。根据研究，油田水中脂肪酸含量可高达2000 ppm以上。按本成油理论，低分子烃类系水溶液中的脂肪酸，由微生物脱羧直接形成。因此，与石油伴生的油田水必然含有较多的脂肪酸。

（2）原生封闭油田水呈碱性，pH > 7。

（3）地球化学勘探方法 $\Delta C$ 的 $CO_2$ 的形成机制。

（4）石油储层次生孔隙的形成。

上述(2)、(3)、(4)三种伴生迹象，均可以用本理论的同一机制解释。因为，石油烃类是由脂肪酸脱羧形成的，即形成烃类的同时，形成二氧化碳。一分子烃类对应一分子的二氧化碳。所以，二氧化碳与石油的关系极为密切，而且二氧化碳的储量极为丰富。二氧化碳与围岩或胶结物中的碳酸盐矿物在有水存在的条件下发生反应，使得水呈碱性，并使储层中碳酸盐胶结物溶解，形成次生孔隙；形成的二氧化碳必定向上部或围岩扩散（或通过其水的作用进入上部的沉积物、地表土壤等），造成 $\Delta C$ 这种侵蚀性的二氧化碳与油藏的紧密相依的关系。

## 2 结论

本理论对油气生成的时间、地点、方式以及条件做如下论述：

生油前期——准备阶段

有机物（包括陆源的与水生的）随同各类碎屑物沉积后，在未失去游离水前，受到生物化学作用和各种分异作用的影响，重新分配，最轻的细分散状的以及胶体状的有机物聚集在细粒沉积物（如泥质）的表面，并和上部继续沉积覆盖的粗碎屑（如砂粒）相混，从而处于含水的多孔介质之中。这种孔隙中的有机物质在继续进行微生物“发酵”作用的时候，沉积覆盖了另一个沉积旋回的细粒沉积物，此种沉积物作为盖层，从而形成一个隔氧的饱含水的封闭体系。

在此饱含水的封闭体系中，有机物质经微生物的长期彻底改造，形成各种脂肪酸类、类石油胶质、类石油沥青质以及甲烷等。此阶段类似于人工沼气发酵。这个阶段无生油作用发生，仅是一种生油前期的准备阶段，亦可称为形成甲烷与脂肪酸阶段。

成油——聚集阶段

在饱含水的封闭体系中，继形成脂肪酸的生化反应之后，开始了脂肪酸的脱羧生物化学反应。这种反应形成各种气态烃类以及各种高分子烃类。这些烃类，加上胶质与沥青质，构成了早期生成的石油。这个阶段相当于碳酸盐岩的固结成岩期。

随着埋深作用增强及气态烃、可汽化的烃类等气体的生成，体系中的压力随之增大，温度增高。由于压力与温度的作用，分散的石油聚集成体积大小不等、含油级别不同的含油体。温度的另一种作用是类石油胶质及类石油沥青质降解形成一部分烃类、胶质与沥青质。这部分烃类与生物作用（即脂肪酸脱羧）形成的烃类构成了封闭体系中全部石油烃类之总和。

在高渗透性的饱含水的封闭体系中，所形成的气态烃（伴生气）、液态烃、胶质及沥青质在压力的作用下，气态烃进一步液化，石油各组分分子间内聚力增强、相互溶解性增强、黏度降低、运移能力增强、分散状的石油聚集成为一个整体，并受岩性等条件的影响，最终形成规模不等的油藏。

（收稿日期：1986 年 11 月 5 日）

**参考文献**

[1] Brassell SC, McEvoy J, Hosfmann CF, et al. Isomerisation, rearrangement and aromatisation of steroids in distinguishing early stages of diagenesis[J]. Org Geochem, 1984,6: 11 - 23.

[2] 何志高. 松辽及其他陆相盆地生油岩地球化学基本特征[J]. 石油实验地质, 1981,3(3):197 - 206.

[3] Yoland Califert, Jean-Louis Qudin et Bernard van de weide. Evolution experimental d'huiles brutes et de fractions d'huiles brutes sous l'influence de la temperature de la pression et de mineraux[J]. Advances Org Geochem, 1968.

[4] Duchscherer W. Geochemical methods of prospecting for hydrocarbons[J]. Oil Gas J, 1980,78:194 - 204.

[5] 张朴瑚. 克拉玛依上二迭统组碎屑储集体中自生沸石的析出与孔隙演化[M]. 北京:地质出版社,1984.

本文原载于《石油实验地质》1988 年第 4 期

# 2. A model of oil-generation in a waterlogged and closed system

He Zhigao

(The Central Laboratory of Petroleum Geology, Ministry of Geology and Mineral Resources, Wuxi, Jiangsu, People's Republic of China, 214151)

**Abstract**: This paper presents a new model on synthetic effects on oil-generation in a waterlogged and closed system. It is suggested based on information about oil in high pressure layers (including gas dissolved in oil), marsh gas and its fermentative solution, fermentation processes and mechanisms, gaseous hydrocarbons of carbonate rocks by acid treatment, oil-field water, recent and ancient sediments, and simulation experiments of artificial marsh gas and biological action.

The model differs completely from the theory of oil-generation by thermal degradation of kerogen but stresses the synthetic effects of oil-generation in special waterlogged and closed geological systems, the importance of pressure in oil-forming processes, and direct oil generation by micro-organisms.

Oil generated directly by micro-organisms is a particular biochemical reaction. Another feature of this model is that generation, migration and accumulation of petroleum are considered as a whole.

## Introduction

This model of oil-generation divides the whole process of oil formation into two stages, preparation and generation-accumulation. For the preparation stage, evidence will be drawn from two sources: firstly from the simulation experiment of the redistribution of organic matters by bio-action and physical differentiation in fine clastic sediments; secondly the comparison of information on the isotopes and components of marsh gas and methane in petroleum gas, which indicates that oil is not formed by biochemical reaction at this stage. For the generation-accumulation stage, seven lines of evidence will be discussed below.

*Redistribution of organic matters in fine clastic sediments*

A simulation experiment on bio-action and physical differentiation was done. It was found that organic matters in sediments could be redistributed, which resulted in redistribution of organic matter deposited in an aqueous medium (Fig. 1). Light and fine organic matters floated and settled on the top portion, together with colloid and colloidal materials which accumulated on the surface of the sediment.

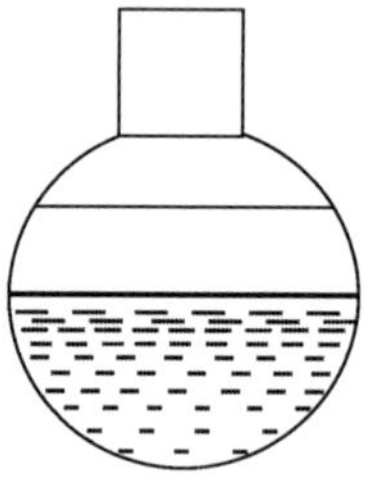


**Fig. 1. Simulation experiment on the redistribution of organic matters**

*Non-production of petroleum hydrocarbons by biochemical reaction in the preparation stage*

There is no doubt that in an aqueous medium, finely dispersed organic matters perform biochemical reactions as indicated by previous studies. Brassell (1984) pointed out that in the water layer and surface of sediment, microaction is the main reaction which plays an important role in steroid isomerization, structural rearrangement and aromatization. From the study of marsh gas fermentation, it is well known that bio-action mainly produces methane and aliphatic acids without petroleum hydrocarbons in the initial stage.

Table 1 shows the results of a study on carbon isotopes of methane in artificial marsh gas with increasing days of fermentation and also in natural gas. With increasing fermentation time, the $\delta^{13}C‰$ increases especially for $CO_2$. When fermentation was carried out for 19 days, the carbon isotope of methane did not reach the range of carbon isotope of methane in petroleum, and the same too for carbon isotopes of $CO_2$ and $CO_2/CH_4$. The gas by fermentation did not contain significant ethane or other heavier alkanes. Therefore, it is believed that no petroleum hydrocarbons form at this stage.

**Table 1. Comparison of data of marsh and natural gas (Marsh gas from Institution of Marsh Gas of Jiangsu, China. Jiangsu natural gas from Tertiary oil field north of Jiangsu, Isotope data from MS analysis)**

| Fermentation time (d) | $CH_4$ $\delta^{13}C$‰ (PDB) | $CH_2$ $\delta^{13}C$‰ (PDB) | $\frac{CO_2}{CH_4}$ (V/V) | $\frac{C_2^+}{\Sigma HC}\times 100$ (%) | $CO_2$ $\delta^{18}O$‰ (PDB) |
|---|---|---|---|---|---|
| 2 | -53.48 | -13.47 | 0.56 | <0.001 | -4.88 |
| 5 | -54.69 | -10.45 | 0.70 | <0.001 | -4.85 |
| 7 | -53.95 | -8.51 | 0.87 | <0.001 | -4.79 |
| 9 | -53.33 | -7.85 | 1.14 | <0.001 | -4.63 |
| 13 | -52.92 | -7.50 | 1.19 | <0.001 | -4.69 |
| 15 | -52.75 | -8.20 | 1.22 | <0.001 | -4.88 |
| 17 | -52.13 | -8.12 | 1.17 | <0.001 | -4.39 |
| 19 | -48.01 | -6.77 | 0.93 | <0.001 | -4.51 |
| Natural gas in Jiangsu | -30.50 | -5.14 | 0.01 | 32 | -3.28 |

*Petroleum hydrocarbon gases originating from decarboxylation of aliphatic acids by organisms*

Aliphatic acids are one of the products of biochemical reactions, a fact which was not only determined by many workers, but was also confirmed by the analysis of fermentation liquid in marsh gas. In a waterlogged and closed system, $C_1 - C_4$ acids can dissolve in water completely and $C_5 - C_7$ acids have high solubility in water; the dissolution amount of $C_8 - C_{10}$ acids in water cannot be underestimated owing to a great amount of water in this system. The geochemical behaviour of these acids in the closed system are described in the following cases:

(1) *The low possibility of thermodecomposition.* According to the chemical properties of aliphatic acids, low aliphatics are not decomposable. When distilled under normal pressure at the boiling point, they are stable. The boiling points of acetic acid and butyric acid are 118 ℃ and 163 ℃ respectively. Hence, within the temperature range in which petroleum usually exists, the thermodecomposition and thermodecarboxylation of aliphatic acids would not take place. This conclusion is supported by the fact that aliphatic acids have been detected in crude oil, oil-field water and ancient rocks at depth.

In addition, a large number of geochemical studies have indicated that with an increase of burial, there is a tendency for the distribution of odd-even carbon of aliphatic acids in rocks and odd-even hydrocarbons in petroleum being inconsistent. If the hypothesis of hydrocarbons being derived from thermodecarboxylation of aliphatic acids is right, then the odd carbon prominence in hydrocarbons will certainly increase (or the even carbon prominence hydrocarbons will decrease) with an increase in burial. In fact, in geological sections with increasing depth, the odd carbon of hydrocarbons in fact decreases, or even carbon increases (except for special environments such as gypseous salt), and tend towards an odd/even value of 1. Therefore, this hypothesis is contradictory to the fact and it is concluded that thermodecarboxylation of aliphatic acids would not yield hydrocarbons.

(2) *Less possibility of biodegradation.* Biodegradation of aliphatic acids commonly occurs in the fermentation liquid of marsh gas (higher molecular aliphatic acids regularly degrade into lower, Bryant 1979) and one would expect that enhancement of this action would lead to a higher methane yield.

In a waterlogged and closed system, although the occurrence of biodegradation cannot be denied, the chance of forming free energy of degradation ($\Delta G$) positively increases owing to the increase of hydrogen partial pressure. Degradation is, however, difficult to proceed (Fig. 2) as this hydrogen effect increases with an increase in aliphatic carbon number. Therefore, it is suggested that biodegradation is not the main form of change for the aliphatic acids. In view of the abundant heavy hydrocarbon composition in petroleum, this action is not considered to have taken place.

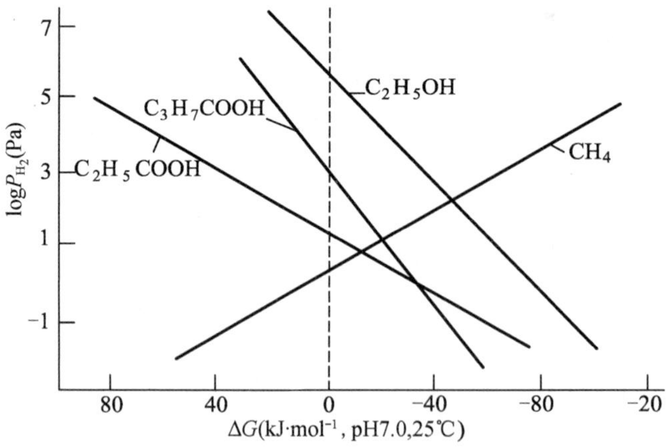


**Fig. 2. The component pressure of $H_2$ influencing Gibbs free energy change($\Delta G$) of degradation of ethanoic acid and forming methane**

(3) *Biological decarboxylation to form hydrocarbons*. The author found that when aliphatic acids in marsh gas (from Institution of Marsh Gas, Chendo) fermentation liquid are treated by special isomer regulation to decarboxylate forming hydrocarbons, their components have a good correlation with the components of petroleum gas. This correlation is shown by the good agreement of the curve patterns of the contents of petroleum gas and aliphatic acids (Fig. 3). They have a very good linear correlation coefficient of 0.9935.

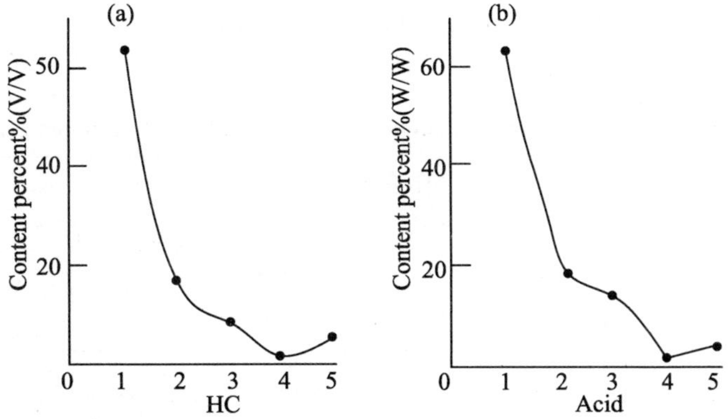


**Fig. 3.** **(a) Distribution of dissolved gaseous hydrocarbons.** 1. **$CH_4$**; 2. **$C_2H_5$**; 3. **$C_3H_8$** 4. **iso-$C_4H_{10}$**; 5. **$nC_4H_{10}$**. **(b) Distribution of normal fatty acids by isomer regulation.** 1. **$CH_3COOH$**; 2. **$CH_3CH_2COOH$**; 3. **iso-butyric and *n*-butyricacid**; 4. **$\frac{1}{2}$iso-pentanoicacid**; 5. ***n*-pentanoic and $\frac{1}{2}$iso-pentanoicacid**

Isomer regulation of aliphatic acids by decarboxylation to form hydrocarbons is not only explained theoretically, but is also tested by its correlation. Regulation by decarboxylation is as follows:

(1) *n*-butanoic acid (iso-butanoic acid)→propane

$$CH_3 \cdot CH_2 \cdot CH_2 \cdot COOH \rightarrow C_3H_8 + CO_2$$

$$\begin{array}{l} CH_3 \cdot CH \cdot COOH \rightarrow C_3H_8 + CO_2 \\ \qquad\quad \bullet \\ \qquad\; CH_3 \end{array}$$

(2) 2-methyl butanoic acid (*n*-pentanoic acid)→*n*-butane

$$\begin{array}{l} CH_3 \cdot CH_2 \cdot CH \cdot COOH \rightarrow nC_4H_{10} + CO_2 \\ \qquad\qquad\quad\; \bullet \\ \qquad\qquad\; CH_3 \end{array}$$

$$CH_3(CH_2)_3 \cdot COOH \rightarrow nC_4H_{10} + CO_2$$

(3) 3-methyl butanoic acid→*i*-butane

$$\begin{array}{l} CH_3 \cdot \underset{\displaystyle CH_3}{\underset{\bullet}{CH}} \cdot CH_2 \cdot COOH \rightarrow CH_3 \cdot \underset{\displaystyle CH_3}{\underset{\bullet}{CH}} \cdot CH_3 + CO_2 \end{array}$$

The thermodynamic properties of the decarboxylation reaction of aliphatic acids to form hydrocarbons such as ethanoic acid, are analysed in Table 2. From Table 2, it can be seen that decarboxylation of ethanoic acid is an exothermic reaction, entropy increasing and free energy decreasing under a thermodynamic standard state. In a waterlogged and closed system, temperature and pressure have less influence on the thermodynamic state functions mentioned above. They also do not lead to a change in the reaction tendency.

**Table 2. Thermodynamic data of decarboxylation of ethanoic acid**

| Parameter | $CH_3$—COOH(l) ⟶ | $CH_4$(g) + | $CO_2$(g) | Parameter change value |
|---|---|---|---|---|
| $\Delta Hf^0$ | -487.3 | -74.9 | -393.8 | $\Delta H = -18.7$ kJ · $mol^{-1}$ |
| $S^0$ | 159.9 | 186.3 | 213.8 | $\Delta S = 240.1$ J · $deg^{-1}$ · $mol^{-1}$ |
| $\Delta G_f^0$ | -392.7 | -50.87 | -394.7 | $\Delta G = -52.8$ kJ · $mol^{-1}$ |

According to CRC thermodynamic regulation and relative information, we can calculate the Gibbs free energy changes of propanoic and other acids. All of these values are about $-41$ kJ · $mol^{-1}$ (for whole reaction kJ/reaction) and located in energy range of organisms physiological reaction (32 – 50 kJ · $mol^{-1}$).

*Hydrocarbon-generation during diagenesis*

The study of carbonate rocks by He (1986) indicates that the hydrocarbon released from carbonate rocks by acid decomposition is primary hydrocarbon included in minerals formed during diagenesis of the carbonate rocks. These hydrocarbons are shown in Table 3 and they can be considered to be petroleum hydrocarbons from their composition, isomer and carbon isotope ratio. From this we think that the time of formation of the petroleum gaseous hydrocarbons is simultaneous with the consolidation-diagenesis phase of the carbonate rocks.

**Table 3. Comparison of data of gas released from Triassic carbonate rocks from Pinba, Guizhou by acid treatment and oil-field gas**

| Samples \ Item | $CH_4$ $\delta^{13}C$‰ (PDB) | $C_2H_6$ $\delta^{13}C$‰ (PDB) | $C_3H_8$ $\delta^{13}C$‰ (PBD) | $C_4H_{10}$ $\delta^{13}C$‰ (PDB) | $\frac{nC_4}{iC_4}$ (V/V) | $\frac{C_2^+}{\Sigma HC}\times 100$ (%) |
|---|---|---|---|---|---|---|
| Released gas(core) | −53.05 | −28.17 | −26.57 | −26.09 | 1.13 | 22.1 |
| Released gas(core) | −34.66 | −26.44 | −24.35 | | | |
| Jiangsu oil-field gas Ed | −46.28 | −32.80 | −29.41 | −28.75 | 2.8 | 32.2 |
| Jiangsu nature gas Ef | −37.93 | −27.81 | −26.22 | −26.43 | 2.5 | 9.9 |

*Pressure effect*

It is considered that pressure plays an important role in the migration and accumulation of petroleum, based on numerous geological phenomenon and a comparison of the properties of crude oil in high pressure strata with that of surficial crude oil. Figure 4 shows a core sample from a depth of 800 m in a well in the Silangmiao structure in the southern margin of the Eerduosi Basin, China (siltstone mainly containing siliceous cement, effective porosity about 10%, absolute permeability 0.1 mD). Figure 4 (earlier) shows a section of the fresh core in which the fissures are filled with crude oil but without any petroleum in the surrounding rock or fissures. Figure 4 (later) shows a section of a core similar to the earlier core but found scattered about the soil around the well for several years. The section of the core occurred in the surficial area has been found to have the surroundings of the fissures and pores of the rock filled by oil.

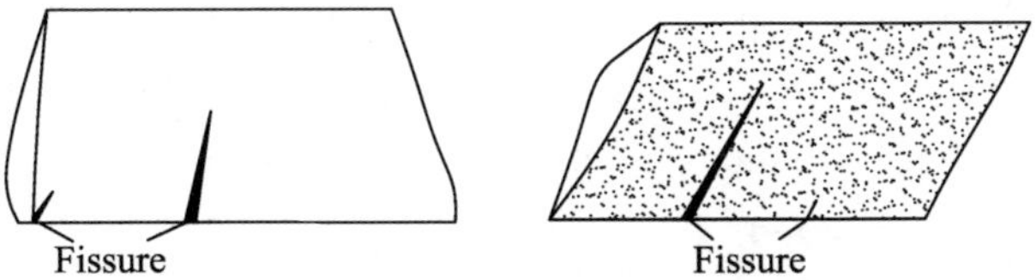


**Fig. 4. Pressure effects displayed by crude oil in fissures**

This phenomenon is commonly present. The author thinks this phenomenon may indicate that pressure can result in various types of distribution of the petroleum in the core samples. When the pressure is high, dispersed petroleum accumulates and moves into channels or fissures in rocks and the petroleum molecules are strongly associated with each other due to the rise in the associated intermolecular force. Therefore, it will be difficult for oil to permeate into the country rock which has very low permeability. When the pressure drops, petroleum disseminates into the country rock which has unfavourable physical properties. It is the former which decreases the accumulation effect, and it is the latter which results in the dispersion of petroleum in the accumulative state.

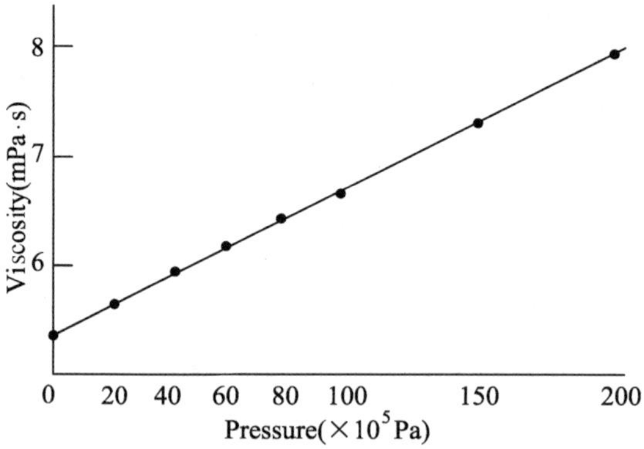


**Fig. 5. The relation between pressure and viscosity**

The relationship between crude oil viscosity and pressure has been determined and is shown in Fig. 5. As is well known, viscosity is a measure of the intermolecular friction force and its values indicate the strength of the intermolecular force. It is obvious that as the distance between molecules becomes nearer, the force will increase with increasing pressure. So viscosity as a measure of internal friction certainly rises with increasing pressure. With respect to non-hydrocarbon-rich crude oil, polar molecules can produce induced dipoles resulting in greater intermolecular forces than that of crude oil of high evolutional grade. Moreover, pressure in particular initially has more influence on the accumulation of dispersed heavy oil generated.

On the other hand, the degree of liquefaction of low molecular hydrocarbons increases with increasing pressure, which benefits the migration and the scale of the accumulation of petroleum.

*Heat effect*

With increase of sediment burial, temperature rises and thermal effects increased. The three main forms of thermal effects are: (1) Thermodegradation will increase the amount of hydrocarbons. Numerous data have already shown that for crude oil itself or extracts from rock, the hydrocarbons present will increase with increase in depth. Beyond all doubt, this thermodegradation commonly exists in a closed system. (2) Due to the increase in temperature, the gasification of hydrocarbon increases and gases expand to lead to an increase in pressure. (3) Petroleum viscosity will drop with increase in temperature. Without doubt, this effect would favour the migration of petroleum.

*Example of initially formed waterlogged and closed system*

Studies of closely spaced samples of Quaternary siltstone beneath recent sediments from a well in offshore China had located the presence of muddy siltstone under sludge which had initially formed a waterlogged and closed system in which petroleum had generated. This kind of petroleum exists from 0 to 1 m as shown in Table 4.

**Table 4. Typical example of oil-generation in a closed system (Samples from ocean drilling, South of Qingdao, China)**

| Location of sample(m) | Rock property | Extract fraction | | | | IR of Aromatic HC |
|---|---|---|---|---|---|---|
| | | Saturated HC(%) | Aro-HC (%) | Resin (%) | Aspha- (%) | $\frac{A_{810}}{A_{750}}$ |
| 0 - 1 | Muddy siltstone | 52.0 | 8.8 | 37.5 | 3.5 | 0.82 |
| 1 - 2 | Muddy siltstone | 24.0 | 5.5 | 62.5 | 7.9 | 0.70 |
| 4 - 5 | Muddy siltstone | 9.7 | 4.9 | 75.7 | 9.8 | 0.50 |

Table 4 shows that extracts from 0 to 1 m sediment samples from the muddy siltstone is indeed petroleum. Towards the bottom, the component of petroleum decreases. This change is believed to be caused by an upward migration of petroleum in the early stages. From this, we can infer that petroleum is generated during diagenesis in a waterlogged and closed system. Also the early migration of petroleum and subsequent accumulation of the petroleum occurred as soon as did the petroleum formation.

*Applications of the model*

The model of oil generated by synthetic effects in a waterlogged and closed system can be applied to solve numerous problems for petrogeology. Some examples are briefly given below:

(1) Numerous "immature" crude oils and oil-fields which are without any known correlative source rocks can be reasonably explained. Because the model provides for the migration and accumulation of petroleum as soon as it is formed and as there are abundant colloidal matter and asphaltene in them, they therefore behave as "immature". The model does not postulate that after the petroleum is formed, they will migrate from the "source rock" to a reservoir rock. Hence, there is no necessity for "organism-rich compact source rocks".

(2) In numerous geochemical analyses of clastic rocks in oil-fields, the results appear to support this model. For all sandy sediments or mudstones containing sand, the potential for oil-generation is higher. The data from the Biyang oil-field in China, especially, provide strong evidence. The cause of this feature is attributed to the fact that sandy sediments which have greater porosity than muddy sediments contain more water and so reworking of micro-organisms to deposit organic matters can proceed for a longer time and to a greater degree so oil-generation is more complete and the potential for oil-generation is higher (He 1981).

(3) In carbonate areas, the zone of facies change between clastic and carbonate sediments has high porosity and contains more water. Hence oil-generation is stronger and this system certainly would have rich petroleum potential.

Generally, all problems which are difficult to be explained by the theory of oil-generation by thermodegradation of kerogen, such as the origin of oil in the oil-field of North Jiangsu, the petroleum reserve of several oil-fields in Ji Zhong, China, can be reasonably well explained, by an application of the model set out in this paper.

*Features related to the formation of an oil-pool*

When an oil-pool forms, the lithological character, property of water and the surrounding medium will change, both physically and chemically. It is a test for the theory on the origin of oil and gas that these changes can be explained correctly.

This model will explain the following associated features:

(1) Higher content of aliphatic acids in oil-field water.

(2) Original oil-field water in alkaline, pH $>7$.

(3) The formation mechanism for CO, found in soil geochemical exploration ΔC (*the soil carbonates had been altered by hydrocarbon leakage form the underlying petroleum, Duchscherer* 1980).

(4) The formation of a secondary pore in a petroleum reservoir.

For feature (1) data available show that the content of aliphatic acids in oil-field water can reach up to 2000 ppm. The model as set out in this paper considers that the aliphatic acids in aqueous solution of low molecular hydrocarbons is derived from the decarboxylation of micro-organisms. Hence the oil-field water associated with petroleum has a higher content of aliphatic acids.

The three associating features (2), (3) and (4) mentioned above can be explained by the mechanism explained by this new model. Because petroleum hydrocarbons result from decarboxylation of aliphatic acids, carbon dioxide is formed at the same time as hydrocarbons. One hydrocarbon will correlate with one $CO_2$. Both petroleum and carbon dioxide are closely related and the amount of $CO_2$ is very high.

Carbon dioxide reacts with carbonate minerals in the surrounding rocks or cement in the presence of water, which makes the water alkaline. When the carbonate cement in the reservoir dissolves, secondary pores will form. Carbon dioxide that has been formed disperses upwards or towards the surrounding rocks (or enters into the sediments and the surficial soil above), which results in the close relationship between carbon dioxide of this kind and the oil-pool.

## Conclusions

The model presented in this paper deals with the time, place, mode and condition for oil and gas formation.

*Prior to oil-generation—preparation stage*

After organic matters (including terrestrial and water-derived sources) were deposited with clastic sediments, and before free water in the sediments is lost, the organic matter will be redistributed, under the influence of biochemistry and differentiation, in a waterlogged and porous environment. The lightest, finest colloidal organic matters will then be dispersed and will accumulate on the surface of fine sediments (e. g. mud) and will be later mixed with coarser sediments (e. g. sand). The organic matters in the waterlogged pores will continuously be fermented by bacteria. Therefore a waterlogged and closed system is formed which is devoid of oxygen.

The organic matters in the system will be completely transformed over a long time into different types of aliphatic acids, oil-like colloids, bitumen, methane, etc. This stage is similar to marsh gas formation, but is not identical. As no oil is generated at this stage, it is just a preparation period and could be called the formation stage of methane and aliphatic acids.

*Oil-generation—accumulation stage*

After the cessation of biochemical reactions responsible for forming the aliphatic acids, the biochemical decarboxylation of aliphatic acids will begin, which will produce various gaseous and heavy hydrocarbons. These hydrocarbons, colloids and asphaltenes consist of early generated petroleum. This stage corresponds to that of the solidification of carbonate rocks.

As the pressure and temperature rise, with increasing burial, gaseous and gasifiable hydrocarbons will be formed, dispersed and accumulated in oil-bearing bodies by the effects of pressure and temperature. Another result of the rising temperature is the conversion of a part of the oil-like colloids and asphaltenes into hydrocarbons, colloids and asphaltenes. These hydrocarbons and biogenetic products (produced by degradation of aliphatic acids) will provide all the hydrocarbons in petroleum.

In a highly permeated and closed system where abundant water is present, the gaseous and liquid hydrocarbons, resins and asphaltenes are formed. Under the effects of pressure the gaseous hydrocarbons are further liquified and the cohesive force among the molecules of the oil fractions will increase; the solubility of the molecules will also increase and so will the ability to migrate. The migrating oils will accumulate to form oil pools of different sizes according to the characteristics of the reservoirs.

## References

[1] Brassell SC, McEvoy J, Hosfmann CF, et al. Isomerisation, rearrangement and aromatisation of steroids in distinguishing early stages of diagenesis[J]. Org Geochem, 1984, 6: 11-23.

[2] Bryant MP. Microbial methane production—theoretical aspects[J]. J Anim Sci,

1979, 48(1): 193 -201.

[3] Duchscherer. Geochemical methods of prospecting for hydrocarbons[J]. Oil Gas J, 1980, 78: 194 -204.

[4] He Zhigao. The basic geochemical characteristic of the source rocks in Songliao and other basins of non-marine facies[J]. Expl Petrol Geol J, 1981, 3(3): 197 -206.

[5] He Zhigao. On the origin of petroleum hydrocarbon gases[J]. Acta Geologica Sinica, 1986, 60: 57 -65.

本文原载于

*Journal of Southeast Asian Earth Sciences*

Vol.5, Nos 1 -4, PP.465 -469, 1991

# 3. 论石油烃类气体成因*

## On the origin of petroleum hydrocarbon gases

何志高

(地质矿产部石油地质中心实验室,江苏无锡,214151)

**摘要:** 本文对中国苏北油田的高压地层原油中油溶气态烃类的组成因素进行了分析。对沼气发酵液中低分子挥发性脂肪酸脱羧所形成的烃类气体组成做了对比;对某些反应式进行了热力学计算,对甲烷和二氧化碳的碳同位素组成进行了分析对比。最后提出石油烃类气体不是通过热解或干酪根热降解形成,而是经生物脱羧反应形成的。这种生物反应,不是目前人们所进行的人工发酵沼气的生物化学反应,而是在沼气发酵以后所进行的生物化学反应。

与液态石油共存的气态烃类的成因和石油成因一样,亦是长期以来研究者所争论的复杂问题。近20年来,在研究石油正构烷烃系列化合物成因时,研究了近代和古代沉积物中的脂肪酸,并做了大量模拟实验,得出了石油正构烷烃可能形成于正构脂肪酸脱羧反应的推论。在这些工作中,主要是研究较高分子($C_9$—$C_{34}$)的脂肪酸,迄今很少涉及低分子挥发性脂肪酸及其低分子气态烃类[1]。

本文报道沼气发酵液中的脂肪酸与高压地层原油中油溶气态烃类在组成上高度的相关性,提出石油类天然气成因的证据。并明确地指出,石油天然气(包括甲烷至丁烷气)是低分子挥发性脂肪酸脱羧而形成的,而非目前流行的石油成因理论,即由生油岩中的高分子的干酪根热降解形成。毫无疑问,这些资料对于石油成因的研究以及各种类型天然气成因的研究均有裨益。

---

* 本文1985年5月收到,1986年3月改回,胡亚昌编辑。

## 1 石油类气体与发酵液脂肪酸组成

人工沼气发酵液中挥发性脂肪酸，若按本文提出的异构体规则处理，则脱羧形成的烃类的组成与高压地层原油油溶气的组成之间存在着高度的线性相关性。这是一种新的发现。

### 1.1 高压地层原油油溶气及其组成的变化

在地质体中，通常与石油有关的烃类气体的组成复杂多变，尤其受到多种来源的干扰和赋存状态的影响。然而，溶解在原油中的气体，是与原油本身在成因上关系最为密切的石油烃类气体。从高压地层原油取样器脱附出来的气体，在消除了压力因素、油藏保存条件优劣以及可能的生物降解等因素后，是最有代表性的石油烃类气体。根据研究，我国苏北油田原油属下第三系原油，演化程度不深，原油组成、性质及地球化学特征比较清楚，决定油溶气组成的因素已经基本掌握。因此，苏北油田及其原油是很合适的研究对象。

表 1 所示是从苏北油田近 40 口井的高压地层原油油溶气的分析资料中所选取的分析项目较全的部分油井资料。表中所列样品和项目具有良好的系统性。

从表 1 可见：随着油层深度的增加，地层压力、油层温度、饱和压力、地下原油与地面原油比重等发生明显的规律性变化；更值得注意的是，随着深度的增加，气体中乙烷含量较丙烷含量愈来愈高。井深在 2174 m 以上的各油井原油，$C_2 < C_3$；井深在 2307 m 以下则 $C_2 > C_3$，且 $C_2/C_3$ 值愈来愈大。

表 2 所示是 CH-22 井同一个高压地层原油样品，分阶段减压、多次分泌，从原油中释放出来的气体的组成。由表 2 可见，油溶气组成的变化与压力因素关系极大。当压力为 $14 \times 10^5$ Pa 时，$C_1 > C_2 > C_3$；当压力为 $7.2 \times 10^5$ Pa 时，$C_1 < C_2 < C_3$。这些资料同表 1 是完全一致的。

**表 1 高压地层原油油溶气态烃综合资料***

**Table 1 Summary of the data of oil-soluble gaseous hydrocarbons in high-pressure in-place oil**

| 井号 | 井深 (m) | 地层 | 地层压力 ($10^5$Pa) | 油层温度 (℃) | 饱和压力 ($10^5$Pa) | 地下原油比重 | 地面原油比重 $D_4^{20}$ | 气体组成(*V/V*) | | | | | | | | |
|---|---|---|---|---|---|---|---|---|---|---|---|---|---|---|---|---|
| | | | | | | | | $C_1$ | $C_2$ | $C_3$ | $iC_4$ | $nC_4$ | $iC_5$ | $nC_5$ | $N_2$ | $CO_2$ |
| 20 | 1910 | Es | 193.1 | 68 | 21.2 | 0.7972 | 0.8405 | 41.1 | 15.2 | 20.2 | 3.6 | 10.0 | 2.7 | 2.8 | 4.0 | 0.5 |
| 12 | 1926 | Es | 197.4 | 70 | 22.4 | 0.7998 | 0.8395 | 45.4 | 16.1 | 20.9 | 3.9 | 8.9 | 2.0 | 1.7 | 0.8 | 0.4 |
| 21 | 1932 | Es | 189.2 | 69 | 16.5 | 0.8041 | 0.8435 | 40.6 | 18.0 | 23.6 | 3.8 | 9.7 | 1.7 | 1.5 | 0.1 | 0.7 |
| 36 | 1953 | Es | 195.3 | 68 | 16.4 | 0.8008 | 0.8404 | 43.1 | 16.8 | 21.4 | 3.4 | 8.7 | 1.5 | 1.5 | 3.3 | 0.3 |
| 49 | 1956 | Es | 189.7 | 70 | 18.0 | 0.7943 | 0.8409 | 42.9 | 16.4 | 20.9 | 3.9 | 9.2 | 2.0 | 2.3 | 2.4 | 0.1 |
| 16 | 1962 | Es | 190.5 | 68 | 16.7 | 0.7990 | 0.8413 | 76.6 | 6.1 | 7.9 | 0.6 | 3.7 | 0.8 | 0.9 | 3.2 | 0.2 |
| 20 | 1988 | Es | 186.8 | 73 | 31.3 | 0.7785 | 0.8430 | 42.4 | 16.4 | 18.5 | 3.3 | 9.7 | 2.2 | 3.4 | 1.7 | 0.5 |
| 26 | 2174 | Ed | 221.3 | 71 | 63.8 | 0.7723 | 0.8336 | 63.6 | 11.7 | 11.8 | 8.2 | | 4.3 | | 0.01 | 0.4 |
| 6 | 2197 | Ed | 212.1 | 76 | 60 | 0.7623 | 0.8349 | 57.6 | 13.2 | 14.1 | 2.3 | 7.2 | 2.2 | 2.9 | 0.3 | 0.3 |
| 20 | 2221 | Ed | 216.4 | 72 | 65.5 | 0.7687 | 0.8332 | 65.1 | 7.8 | 11.4 | 2.1 | 6.4 | 1.9 | 2.7 | 1.7 | 0.8 |
| 35 | 2299 | Ed | 230.7 | 83 | 63.5 | 0.7502 | 0.8329 | 51.1 | 15.7 | 15.4 | 2.6 | 7.0 | 1.8 | 6.1 | 0.7 | 0.3 |
| 19 | 2307 | Ed | 215.3 | 78 | | 0.7543 | 0.8335 | 60.4 | 12.9 | 12.7 | 1.7 | 6.6 | 1.7 | 2.4 | 1.2 | 0.4 |
| 48 | 2871 | Ed | 276.8 | 84 | 90.8 | 0.7396 | 0.8264 | 69.2 | 13.0 | 9.4 | 1.2 | 4.2 | 1.4 | 0.6 | | |
| S-80 | 3061 | Ed | 344 | 100 | 160.4 | 0.6870 | 0.8147 | 66.9 | 14.2 | 9.3 | 1.3 | 4.4 | 1.1 | 2.0 | 0.2 | 0.7 |

*样品由江苏石油勘探开发研究所分析。

**表 2 CH-22 地层原油分级采样分析**

**Table 2 Analytic data of in-place oil sampled from well-CH-22 with step-down pressures**

| 压力 | 气体组成(V/V) | | | | | | | | |
|---|---|---|---|---|---|---|---|---|---|
| ($1\times10^5$ Pa) | $C_1$ | $C_2$ | $C_3$ | $iC_4$ | $nC_4$ | $iC_5$ | $nC_5$ | $CO_2$ | $N_2$ |
| 120.8 | 78.7 | 9.5 | 5.0 | 0.5 | 1.3 | 0.2 | 0.4 | 1.0 | 3.3 |
| 100 | 78.2 | 10.1 | 5.2 | 0.6 | 1.5 | 0.3 | 0.4 | 1.0 | 2.7 |
| 76.7 | 75.7 | 12.0 | 6.3 | 0.6 | 1.9 | 0.4 | 0.5 | 0.9 | 1.7 |
| 52.0 | 71.7 | 14.1 | 7.9 | 0.8 | 2.1 | 0.4 | 0.6 | 0.9 | 1.2 |
| 36.8 | 64.4 | 18.0 | 10.6 | 1.0 | 3.0 | 0.5 | 0.7 | 1.0 | 0.4 |
| 24.0 | 52.6 | 22.4 | 15.3 | 1.5 | 4.7 | 0.8 | 1.2 | 1.0 | — |
| 14.0 | 36.7 | 26.4 | 22.6 | 2.5 | 6.6 | 1.3 | 1.8 | 1.0 | — |
| 7.2 | 2.3 | 17.0 | 38.0 | 7.1 | 21.0 | 4.9 | 6.4 | 0.7 | — |

表 3 是苏北原油产层深度及其有关性质、组成变化方面的对比资料。由表 3 可见,原油产层深度与原油正构烷烃色谱分析主峰碳数毫无关系;深度与轻烃( <250 ℃馏分)含量亦无关系,它们之间的相关系数为 -0.0580。这种呈现负值的相关系数表明,原油并未因深度的增加、相应温度的增高而热解成低分子烃类。这种结果与 Yolande Califert[2] 的原油模拟实验资料一致:原油在高温、高压(230 ℃、$300\times10^5$ Pa 72 h)条件下,没有低分子气态烃类产生。因此,表 1 中随原油深度的增加,油溶气中甲烷相对含量增高,不是原油演化程度加深所致,而是由于压力比较高、甲烷在油相中的溶解量较多,或者说是甲烷扩散到非油相的储集空间较少所致。其实,表 1 中的 Es 与 Ed 的分界线上下两套油层(实际上为两套地层的原油)温度只有很小的差别,而甲烷含量则出现了很大的差异,这就可以说明,油溶气中甲烷的量不是由油层温度所决定的。

**表 3 原油产层深度与其性质、组成变化**

**Table 3 Depths of the oil-producing horizons and properties and composition of the crude oil**

| 井深 (m) | 色谱主峰 | $\frac{C_{28+29}}{C_{21+22}}$ | 轻烃 (%) | 饱和烃 (%) | 芳烃 (%) | 胶质 (%) | 沥青质 (%) |
|---|---|---|---|---|---|---|---|
| 1338 | $C_{22}$ | 0.59 | 10.7 | 62.3 | 12.4 | 13.5 | 1.1 |
| 1390 | $C_{22}$ | 0.48 | 18.7 | 61.4 | 8.1 | 9.0 | 2.8 |
| 1500 | $C_{17}$ | 0.80 | 28.5 | 37.2 | 10.7 | 21.9 | 1.7 |

续表

| 井深(m) | 色谱主峰 | $\frac{C_{28+29}}{C_{21+22}}$ | 轻烃(%) | 饱和烃(%) | 芳烃(%) | 胶质(%) | 沥青质(%) |
|---|---|---|---|---|---|---|---|
| 1816 | $C_{22}$ | 0.66 | 8.2 | 70.4 | 10.4 | 9.8 | 1.2 |
| 1819 | $C_{17}$ | 0.68 | 34.3 | 46.9 | 8.5 | 7.0 | 3.5 |
| 2005 | $C_{17}$ | 0.46 | 21.2 | 46.0 | 11.7 | 15.2 | 5.8 |
| 2036 | $C_{22}$ | 0.38 | 10.7 | 63.4 | 6.9 | 17.8 | 1.3 |
| 2064 | $C_{22}$ | 0.43 | 13.5 | 65.5 | 8.0 | 8.4 | 4.7 |
| 2254 | $C_{22}$ | 0.47 | 11.4 | 72.0 | 9.2 | 6.4 | 1.0 |
| 2308 | $C_{22}$ | 0.59 | 35.9 | 50.5 | 7.6 | 4.8 | 1.3 |
| 2432 | $C_{23}$ | 0.72 | 12.2 | 67.8 | 7.3 | 4.7 | 5.7 |
| 2481 | $C_{23}$ | 0.69 | 20.1 | 55.7 | 8.9 | 11.2 | 4.0 |
| 2705 | $C_{15}$ | 0.44 | 15.8 | 69.9 | 6.5 | 5.3 | 1.9 |
| 2859 | $C_{22}$ | 0.63 | 19.5 | 64.6 | 6.6 | 6.0 | 3.3 |
| 3047 | $C_{22}$ | 0.35 | 16.6 | 67.4 | 4.7 | 10.5 | 0.8 |
| 3062 | $C_{22}$ | 0.68 | 16.9 | 62.8 | 8.1 | 8.8 | 3.5 |
| 3102 | $C_{22}$ | 0.76 | | | | | |
| 相关系数 | | 0.319 | -0.058 | 0.349 | -0.681 | -0.396 | 0.128 |

### 1.2 石油烃类气体的代表样品

在上述资料的基础上,根据已述挑选代表性样品的原则,选取了 S-80 井的地层原油油溶气作为石油烃类气体的代表,其组成列于表 1。该样品井深最大,地层压力最高,$C_2 > C_3$,氮含量很低,仅 0.2%,二氧化碳含量亦不高,低于 1%。可以认为,这种气体无空气的混入,未遭受次生氧化作用与微生物作用;该样品甲烷含量较高,但不是原油热降解所致;其地下原油比重很低,主要原因是饱和压力高、溶解气量大,从而说明该样品具有最大限度地保存着原油气体原始组成的特点。因此,该原油中的溶解气是最具有代表性的石油烃类气。其组成由于消除了种种干扰与后生变化的影响,无疑是最接近原始生成状态、最具备原始特征的。

### 1.3 沼气发酵液中脂肪酸的组成

人工沼气发酵液中脂肪酸的组成因发酵物料与发酵条件而异。为了能同所研究的苏北原油对比,取畜粪作为原料的沼气发酵液,其组成如表 4 所示。

**表 4　沼气发酵液中脂肪酸的组成*（*W/W*）**

**Table 4　Composition of fatty acids in the fermentation liquor for marsh gas**

| 脂肪酸名称 | 乙酸 | 丙酸 | 异丁酸 | 正丁酸 | 异戊酸 | 正戊酸 | 丁酸总量 | 1/2 异戊＋正戊酸 |
|---|---|---|---|---|---|---|---|---|
| 含量(%) | 63.6 | 18 | 4.2 | 9.8 | 2.8 | 1.4 | 14 | 2.8 |

* 根据 D. L. Klass 资料整理。

### 1.4　相关性

经过对脂肪酸脱羧形成烃类的异构体规则的研究，发现按异构体规则处理的脂肪酸与油溶气态烃之间存在着密切的相关性。这种相关性主要表现在两者含量变化曲线的形态一致，以及两组数据很大的线性相关系数。

（1）异构体规则

根据 Марковников 理论及一般原理，脂肪酸脱羧形成烃类存在着特定的规则。这些规则的含义是：正丁酸与异丁酸结构虽然不同，但脱羧以后，都生成同样的一种烃类，即丙烷；戊酸具有三种异构体，2-甲基丁酸与正戊酸脱羧后，均生成正丁烷，3-甲基丁酸脱羧后形成异丁烷。这种脱羧模式，不仅在理论上可以做出解释，而且在实际上具有这种关系的定量实例可资对比，即可表示为：

a. 正丁酸＋异丁酸＝丙烷：

$$CH_3 \cdot CH_2 \cdot CH_2 \cdot COOH \longrightarrow C_3H_8 + CO_2$$

$$CH_3 \cdot \underset{\underset{\displaystyle CH_3}{\cdot}}{CH} \cdot COOH \longrightarrow C_3H_8 + CO_2$$

b. 2-甲基丁酸（1/2 异戊酸）＋正戊酸＝正丁烷：

$$CH_3 \cdot CH_2 \cdot \underset{\underset{\displaystyle CH_3}{\cdot}}{CH} \cdot COOH \longrightarrow nC_4H_{10} + CO_2$$

c. 3-甲基丁酸（1/2 异戊酸）＝异丁烷：

$$CH_3 \cdot \underset{\underset{\displaystyle CH_3}{\cdot}}{CH} \cdot CH_2 \cdot COOH \longrightarrow CH_3 \cdot \underset{\underset{\displaystyle CH_3}{\cdot}}{CH} \cdot CH_3 + CO_2$$

（2）极相似的组成变化曲线

按上述规则，脂肪酸脱羧形成对应的烃类，并将这些酸类归一化，求得各单个酸类的百分含量，与表 1 中 S-80 油溶气态烃的单个烃类的百分含量分别作图，发现两者的分布曲线极为相似。如图 1 所示，左图为油溶气态烃组成分布曲线，右图为脂肪酸组成分布曲线。

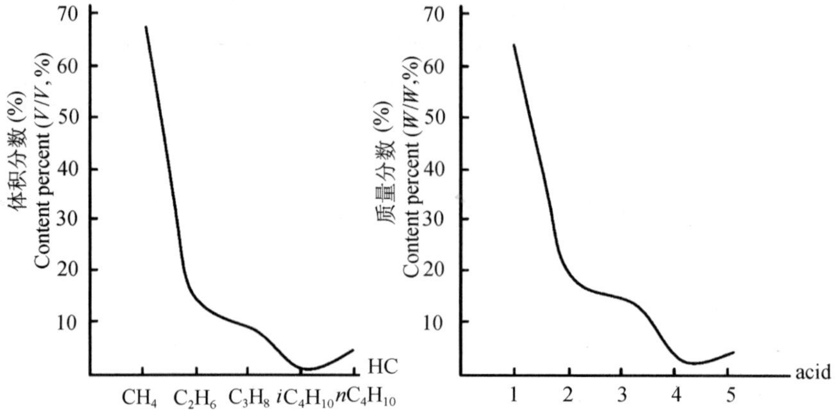


**图 1 油溶气态烃(左)和脂肪酸(右)组成分布**

1. $CH_3COOH$ 2. $CH_3CH_2COOH$ 3. {iso-butyric acid; *n*-butyric acid}

4. 1/2 iso-pentenoic acid 5. {*n*-pentenoic acid; 1/2 iso-pentenoic acid}

**Fig. 1 Distribution of composition of oil-soluble gaseous hydrocarbons (left) and fatty acids (right) in the fermentation liquor**

(3) 高的相关系数值

用数理统计方法处理,表 1 中的 S-80 油溶气态烃百分组成(体积分数),与表 4 中的脂肪酸按脱羧规则处理的脂肪酸百分组成(质量分数)两组数据,其线性相关系数值很高,为 0.9935。由此可见,这两组数据之间的关系密切,两种过程间的相似程度很高。

## 2 讨论

### 2.1 *脂肪酸脱羧反应的热力学分析*

乙酸脱羧形成甲烷,反应的热力学性质可以做如下分析:

$$CH_3\cdot COOH(l) \longrightarrow CH_4(g) + CO_2(g)$$

$\Delta H_f^0$: $-487.0(-116.4)$ $-74.8(-17.89)$ $-393.5(-94.05)$

$S^0$: $159.8$ $(38.2)$ $186.2$ $(44.5)$ $213.6$ $(51.06)$

$\Delta G_f^0$: $-392.5$ $(-93.8)$ $50.8(-12.14)$ $-394.4(-94.26)$

根据计算,此式的 $\Delta H$ 值为 $-18.66\ \mathrm{kJ\cdot mol^{-1}}$ ($-4.461\ \mathrm{kcal\cdot mol^{-1}}$),$\Delta S$ 值为 $240.0\ \mathrm{J\cdot deg^{-1}\cdot mol^{-1}}$(或 $57.36\ \mathrm{cal\cdot deg^{-1}\cdot mol^{-1}}$),$\Delta G$ 值为

$-52.7\ kJ\cdot mol^{-1}$（或 $-12.6\ kcal\cdot mol^{-1}$）。由此可见，乙酸脱羧反应在热力学标准状况下是放热、熵增、自由能降低的反应。

根据CRC规则，求得的丙酸及其更高分子挥发性脂肪酸脱羧反应，其自由能变化值 $\Delta G^0$ 约为 $-41.84\ kJ\cdot mol^{-1}$，均在生物生理变化能量级（$33.5\sim50.2\ kJ\cdot mol^{-1}$）范围[3]。

2.2 脂肪酸及脱羧反应的地球化学分析

脂肪酸广泛分布在原油及古代、近代沉积物中，这是众所周知的。而且，大家都熟悉脂肪酸的偶碳优势的事实。

Kvenvolden等[4,5]研究指出：古代沉积岩中正脂肪酸偶碳数优势小于近代沉积物；石油正构烷烃来自正脂肪酸脱羧作用，由于脱羧而形成比脂肪酸少一个碳数的正烷烃。这就是说，近代沉积物中偶碳数脂肪酸的强优势到了成岩以后的生油岩，脂肪酸脱羧形成比该酸少一个碳的正烷烃，从而出现了脂肪酸偶碳优势的降低。按此推理，正烷烃奇碳优势应该增强。

笔者认为，在脂肪酸的偶、奇优势变化与正烷烃奇、偶优势变化中，显然存在一个与事实不符的矛盾：正烷烃奇数碳优势在岩石中不是如上所述的增强，而是降低；若认为古代岩石中，正烷烃奇数碳优势降低或偶数碳优势得到增加，那么，岩石中的脂肪酸的偶碳数优势不仅不会降低，反而会由于奇碳的减少而比近代沉积物具有更高的偶/奇值（或称酸分布比值），即偶/奇项中分母减小了，其比值当然增大。这显然是与事实不符的。

大量的地球化学研究资料表明，随岩石的埋深作用的增强，正烷烃奇数碳优势与脂肪酸偶碳数优势均在降低。因此，岩石中石油正烷烃的形成，绝不是因比正烷烃高一个碳数的正脂肪酸脱羧——热降解而形成的。也就是说，本文前已论述的异构体规则在岩石中不存在，脂肪酸的那种脱羧过程及其定量关系不存在。

从理论上来看，地温的作用导致脂肪酸的脱羧不会形成奇偶碳数的明显差异。从实际资料来看，脂肪酸不论奇偶碳数，在相当深的钻井中均存在，氨基酸亦是如此，它们都是相当稳定的化合物。而且，古代岩石中存在着诸如碳酸盐岩、膏盐等沉积岩的正烷烃偶碳优势。更值得注意的是，同一样品在不同碳数区间存在不同的碳数优势。这是热解理论难以解释的矛盾与现象。

笔者认为，形成气态烃的脂肪是低分子脂肪酸，它们在水中具有很大的溶解度；固态沉积物（有机物与无机物）对这类水溶性的脂肪酸的吸附是比较小的。事实上，人工沼气发酵液与发酵原料能充分接触，原料对脂肪的吸附已经饱和。因此，形成大量的石油烃类气体的脂肪酸的脱羧反应只能在水体中进行。根据上述热力学分析以及生物作用的特点,脱羧反应很可能是由生物完成的。

表5所示是人工沼气发酵装置——并联反应器不同部位(上、中、下)、不同时间戊酸异构体的含量变化。由表5可见,从戊酸异构体含量的数值与比值来看,随发酵时间的延长,脂肪酸异构体增多,正构体减少,正戊酸/异戊酸减小。如同均以中部为例,比值由2 h的2.44降低到7 h的2.00,再降低到15 h的1.20;其(正戊酸+1/2异戊酸)与1/2异戊酸的比值亦如此。从这种比值的变化来看,要达到脱羧形成的气体,其正丁烷/异丁烷值约为1.5,沼气发酵时间需要大于15 h,这个阶段形成的脂肪酸,其组成方有可能具备形成石油烃类气体的特征。因此,从这个实验资料可见,脂肪酸脱羧形成石油类天然气的阶段是比人工沼气发酵更晚的阶段。

**表5 沼气发酵液中戊酸组成变化与发酵时间、发酵部位的关系**

**Table 5 Changes of the composition of pentanoic acids in the fermentation liquor for methene generation and their relations to the fermentation time and portion**

| 名称 | 正戊酸含量(mg/L) | | | 异戊酸含量(mg/L) | | | 正戊酸/异戊酸 | | | (正戊酸+1/2异戊酸)/(1/2异戊酸) | | |
|---|---|---|---|---|---|---|---|---|---|---|---|---|
| 时间(h) \ 部位 | 下 | 中 | 上 | 下 | 中 | 上 | 下 | 中 | 上 | 下 | 中 | 上 |
| 2 | 157 | 110 | 133 | 49 | 45.1 | 55 | 3.2 | 2.4 | 2.4 | 7.4 | 4.8 | 5.8 |
| 7 | 99.3 | 94.1 | 123 | 41 | 47 | 43.3 | 2.4 | 2.0 | 2.8 | 5.8 | 5.0 | 6.7 |
| 15 | 40.2 | 54.2 | 61 | 50 | 45 | — | 0.8 | 1.2 | | 2.6 | 3.4 | |

发酵条件:反应温度36 ℃~39 ℃,反应器∅=6.2 cm

试验单位:中国科学院成都分院。

表6所示是人工沼气甲烷碳及二氧化碳的碳、氧稳定同位素与发酵时间的关系对比资料。从表6可见,发酵19 d以后,还未见到乙烷等重的气态烃;碳的同位素亦未变到石油类气体的同位素值的范围。因此,石油气的形成不是人工沼气型的生物化学反应,而是在这个阶段结束后的另一种特定类型的生物化学反应。

**表6 沼气、天然气同位素资料对比**

**Table 6 Comparison between the isotope data of methene gas and natural gas**

| 发酵时间 (d) | $CH_4$ $\delta^{13}C$‰ (PDB) | $CO_2$ $\delta^{13}C$‰ (PDB) | $CO_2$ $\delta^{18}O$‰ (PDB) | $\frac{CO_2}{CH_4}$ (体积比) | $C_2^+$ % (体积比) |
|---|---|---|---|---|---|
| 2 | -53.48 | -13.47 | -4.88 | 0.56 | <0.001 |
| 5 | -54.69 | -10.45 | -4.85 | 0.70 | <0.001 |
| 7 | -53.95 | -8.51 | -4.79 | 0.87 | <0.001 |
| 9 | -53.33 | -7.85 | -4.63 | 1.14 | <0.001 |
| 13 | -52.93 | -7.50 | -4.69 | 1.19 | <0.001 |
| 15 | -52.75 | -8.20 | -4.88 | 1.22 | <0.001 |
| 17 | -52.13 | -8.12 | -4.89 | 1.17 | <0.001 |
| 19 | -48.01 | -6.77 | -4.51 | 0.93 | <0.001 |
| 江苏天然气 | -30.5 | -5.14 | | | |

## 3 结论

地质体中石油烃类气体组成复杂多变，主要原因是影响因素很多。在消除了压力以及生物降解等影响因素之后，高压地层原油中油溶气态烃是真正反映石油类气体原始成分的烃类气体。这种气体的组成与沼气发酵液中挥发性脂肪酸按一定规则处理的组成之间有着很强的相关性。其线性相关系数为0.9935，两者的含量变化曲线十分相似。脂肪酸脱羧形成烃类的“异构体规则”这个理论来自于实践，受到实践的检验。

根据脂肪酸由现代沉积物至古代岩石的偶碳优势变小以及正烷烃奇碳优势变小的地球化学事实可以认为，从现代沉积物至生油岩的变化中，不存在本文所提出的脂肪酸脱羧反应的异构体规则，即不存在着偶碳数脂肪酸脱羧生成少一个碳的奇碳数正烷烃的反应过程。因为按异构体规则，脂肪酸偶碳数优势减小，奇碳数正烷烃必定增加。事实是，偶碳数脂肪酸减少，奇碳数正烷烃亦减少（当然，这里不指如膏盐沉积等特殊情况）。因此，在这种前提条件下，按异构体规则脱羧生成烃类的反应不可能是热分解脱羧反应。

低分子脂肪酸由于在水中的溶解度较大，沉积物对它们的吸附力相当小，其脱羧反应必定主要是在水介质条件下进行的。

热力学分析表明，脱羧反应的自由能变化值 $\Delta G^0$ 为 $41.87 \sim 50.2\ kJ \cdot mol^{-1}$，介于生物生理变化能量级的范围。

因此，形成石油烃类气体的脱羧反应应该是生物反应。然而，这种生物反应不是目前人们所进行的人工发酵沼气的生物化学反应。根据沼气发酵液中脂肪酸异构体的变化趋势、沼气中甲烷碳同位素的变化趋势，以及沼气中乙烷以上重烃含量的特征，可以判断，形成石油烃类气体的生物化学反应是在沼气发酵以后所进行的特殊的生物化学反应阶段完成的。

## 参考文献

[1] Seifert WK. Carboxylic acid in petroleum and sediments. Fortschr Chem Org Naturst, 1974, 32:2 -3,38 -39.

[2] Califert Y. Evolution expérimentale d'huiles brutes et de fractions d'huiles brutes sous l'influence de la temperature de la pression et de minéraux [M]// Advances in Organic Geochemisty. Pergamon Press,1968.

[3] Bryant MP. Microbial methane production—theoretical aspects [J]. J Anim Sci, 1979,48(1):193 -201.

[4] Kvenvolden KA. Geochemical Exploration, "United States Patents" 3, 1969, 480, 396, 11.

[5] Kvenvolden KA. Evidence for transformation of normal fatty acid in sediments[M]// Advance in Organic Geochemistry. Pergamon,1966.

本文原载于《地质学报》
1986 年第 60 卷第 4 期

# 4. On the origin of petroleum hydrocarbon gases

He Zhigao

(The Central Laboratory of Petroleum Geology, Ministry of Geology and Mineral Resources, Wuxi, Jiangsu, People's Republic of China, 214151)

**Abstract**: The compositional variation of oil-soluble gaseous hydrocarbons in high-pressure in-place oil from more than 40 oil wells in the Northern Jiangsu oil field, China, was studied. Samples in which the effects of the factors of pressure and biodegradation had been got rid of were chosen as the representatives that could really reflect the original composition of hydrocarbon gases. Such samples were compared compositionally with their corresponding hydrocarbons formed by decarboxylation of volatile low-molecular fatty acids. Thus highly correlative regularities were found to exist between them. The numerical values of both are graphically expressed. The two curves on the diagram are very similar. These two sets of values have a simple correlation coefficient of 0.9935, which presents their genetic relation.

Based on the study of the formation of gaseous hydrocarbons from decarboxylation of volatile fatty acids, the following regularities of the isomers are revealed:

1) *n*-butyric acid + isobutyric acid = propane;

2) 1/2 isopentanoic acid (2-methyl-butyric acid) + *n*-pentanoic acid = *n*-butane;

3) 1/2 isopentanoic acid (3-methyl-butyric acid) = isobutane.

The thermodynamics of decarboxylation was analysed and the enthalpy change $\Delta H$, entropy change $\Delta S$ and Gibbs free energy change, $\Delta G$ in the following reaction were calculated:

$$CH_3COOH(l) \longrightarrow CH_4(g) + CO_2(g)$$

Thus the author considers that the reaction for the formation of hydrocarbons due to the decarboxylation of fatty acids is an exothermic one with increased entropy and decreased free energy as well as a spontaneous one. According to the CRC rule, the resultant Gibbs free energy change $\Delta G$ is approximately $-41.84$ kJ/mol, which is well within the range of the energy levels of biophysiological changes.

According to the changes in odd-carbon preference in $n$-alkanes of petroleum and sediments and the analysis of fatty acids and amino acids, it is considered that hydrocarbon gases do not originate from pyrolysis or thermal degradation but from decarboxylation of organisms. But this biological reaction is not ordinary biochemical reaction in the artificial fermentation for generating marsh gas. On the basis of the change tendency of the fatty acid content in the fermentation liquor for marsh gas generation and the change tendency of the carbon isotopes in methane and carbon dioxide in the marsh gas, it can be judged that the biochemical reaction for forming hydrocarbon gases is a special biological one occurring after the fermentation.

## 1 Introduction

The origin of gaseous hydrocarbons coexisting with liquid oil, just like the origin of oil, has long been a complex problem of controversy. Over the past twenty years, fatty acids in recent and ancient sediments have been studied and many simulation experiments have been done when studying the origin of the homologous series of $n$-alkanes in petroleum. These studies and experiments have led us to the following inference: $n$-alkanes can be produced by decarboxylation of $n$-fatty acids. These studies deal mainly with high-molecular-weight ($C_9$ – $C_{34}$) fatty acids, but seldom with low-molecular-weight volatile fatty acids and gaseous hydrocarbons[1].

This paper reports the high compositional correlation between fatty acids in marsh gas fermentation liquids and oil-soluble gaseous hydrocarbons in high-pressure in-place oil, presents evidence for the origin of petroleum gases, and explicitly points out that petroleum gases (including methane to butane gases) are formed by decarboxylation of low-molecular-weight volatile fatty acids. This idea is different from the present popular theory of oil generation—oil is formed by thermal degradation of high-molecular kerogen in source rocks. Undoubtedly this information is beneficial to the study of oil origin and the origin of various types of natural gases.

## 2 Compositions of Petroleum Gases and Fatty Acids in Fermentation Liquids

According to the rules of isomers proposed in this paper, there is a highly linear correlation between the composition of the hydrocarbons formed by decarboxylation of volatile fatty acids in fermentation liquids of artificial marsh gases and that of oil-soluble gases in high-pressure in-place oil. This is a new discovery.

### 2.1 Oil-soluble gases in high-pressure in-place oil and their compositional variation

In a geologic body, the composition of hydrocarbon gases associated with petroleum is generally complex and variable, and is especially subjected to interferences from multiple sources and influences of their modes of occurrences. Gases dissolved in in-place oil, however, are petroleum hydrocarbon gases that are genetically most intimately associated with in-place oil itself. The gases desorbed from high-pressure in-place oil are the most representative of petroleum hydrocarbon gases when such factors as pressure, preservation conditions of oil reservoirs and possible biodegradation are eliminated. Through studies, we know that the oil from the Northern Jiangsu oilfield, China, belongs to the Lower Tertiary oil with a moderate degree of maturity; its composition, nature and geochemical characteristics are relatively clear, and the factors determining the composition of the oil-soluble gases have been studied in some detail. Thus it is proper to choose the oil of the Northern Jiangsu oilfield as an object of this study.

Table 1 lists the analytical data of the oil-soluble gases in high-pressure in-place oil selected from those of nearly forty wells in the Northern Jiangsu oilfield. The samples and items listed in Table 1 are systematic.

It can be seen from Table 1 that the formation pressure, reservoir temperature, saturation pressure and specific gravities of subsurface and surface oil all show appreciable regular variations with increasing reservoir depth. It is especially noteworthy that the content of ethane in the gas becomes progressively higher than that of protane with the increase of depth: thus in all the oil wells, $C_2 < C_3$ above 2174 m, while $C_2 > C_3$ below 2307 m and the $C_2/C_3$ ratio increases with increasing depth.

Table 2 shows the composition or the gases released from the same high-pressure in-place oil sample raised from Well CH-22 through stepwise pressure decrease and multiple segregation. It can be seen that the variation of the composition of the oil-soluble gases is closely related to the pressure. When the pressure is greater than $14 \times 10^5$ Pa, $C_1 > C_2 > C_3$; when the pressure decreases down to $7.2 \times 10^5$ Pa, $C_1 < C_2 < C_3$. These data are entirely coincident with those in Table 1.

The depths of the oil-producing horizons and the properties and composition of the crude oil in the Northern Jiangsu oilfield are shown in Table 3. It is apparent from the table that the depth of the oil-producing horizons is related neither to the main gas chromatographic peak carbon number of *n*-alkanes of the crude oil nor to the content of light hydrocarbons ( <250 ℃ fraction), their correlation coefficient being $-0.058$. This negative correlation coefficient indicates that the crude oil has not been thermally degraded into high-molecular-weight hydrocarbons with increasing depth and temperature. This analytical result is consistent with that of the simulation experiments of crude oils of Y. Califert (1968)[2]: crude oil does not generate any low-molecular gaseous hydrocarbons even under high-temperature and high-pressure conditions (230 ℃, $300 \times 10^5$ Pa, 72 h). Therefore, the increase of the methane content in the oil-soluble gases with increasing depth in Table 1 is not due to the deepening of the evolution of the crude oil, but due to the increasing amount of methane dissolved in the oil phase under increasing pressure with depth, in other words, little methane has diffused into non-oil phase reservoir space. In Table 1, there is little difference in temperature between two oil reservoirs (actually crude oils of two oil reservoirs) above and below the Es Ed boundary, while there is great difference in methane content. This indicates that the content of methane in oil-soluble gases is not determined by the temperatures of oil reservoir.

**Table 1. Summary of the data of oil-soluble gaseous hydrocarbons in high-pressure in-place oil***

| Well No. | Well depth (m) | Formation | Formation press. ($10^5$ Pa) | Reservoir temp. (℃) | Saturated press. ($10^5$ Pa) | Subsurface oil sp. gr. | Surface oil sp. gr. $D_4^{20}$ | Gas composition (V/V, %) | | | | | | | | |
|---|---|---|---|---|---|---|---|---|---|---|---|---|---|---|---|---|
| | | | | | | | | $C_1$ | $C_2$ | $C_3$ | $iC_4$ | $nC_4$ | $iC_5$ | $nC_5$ | $N_2$ | $CO_2$ |
| 20 | 1910 | Es | 193.1 | 68 | 21.2 | 0.7972 | 0.8405 | 41.1 | 15.2 | 20.2 | 3.6 | 10.0 | 2.7 | 2.8 | 4.0 | 0.5 |
| 12 | 1926 | Es | 197.4 | 70 | 22.4 | 0.7998 | 0.8395 | 45.4 | 16.1 | 20.9 | 3.9 | 8.9 | 2.0 | 1.7 | 0.8 | 0.4 |
| 21 | 1932 | Es | 189.2 | 69 | 16.5 | 0.8041 | 0.8435 | 40.6 | 18.0 | 23.6 | 3.8 | 9.7 | 1.7 | 1.5 | 0.1 | 0.7 |
| 36 | 1953 | Es | 195.3 | 68 | 16.4 | 0.8008 | 0.8404 | 43.1 | 16.8 | 21.4 | 3.4 | 8.7 | 1.5 | 1.5 | 3.3 | 0.3 |
| 49 | 1956 | Es | 189.7 | 70 | 18.0 | 0.7943 | 0.8409 | 42.9 | 16.4 | 20.9 | 3.9 | 9.2 | 2.0 | 2.3 | 2.4 | 0.1 |
| 16 | 1962 | Es | 190.5 | 68 | 16.7 | 0.7990 | 0.8413 | 76.6 | 6.1 | 7.9 | 0.6 | 3.7 | 0.8 | 0.9 | 3.2 | 0.2 |
| 20 | 1988 | Es | 186.8 | 73 | 31.3 | 0.7785 | 0.8430 | 42.4 | 16.4 | 18.5 | 3.3 | 9.7 | 2.2 | 3.4 | 1.7 | 0.5 |
| 26 | 2174 | Ed | 221.3 | 71 | 63.8 | 0.7723 | 0.8336 | 63.6 | 11.7 | 11.8 | 8.2 | | 4.3 | | 0.01 | 0.4 |
| 6 | 2197 | Ed | 212.1 | 76 | 60 | 0.7623 | 0.8349 | 57.6 | 13.2 | 14.1 | 2.3 | 7.2 | 2.2 | 2.9 | 0.3 | 0.3 |
| 20 | 2221 | Ed | 216.4 | 72 | 65.5 | 0.7687 | 0.8332 | 65.1 | 7.8 | 11.4 | 2.1 | 6.4 | 1.9 | 2.7 | 1.7 | 0.8 |
| 35 | 2299 | Ed | 230.7 | 83 | 63.5 | 0.7502 | 0.8329 | 51.1 | 15.7 | 15.4 | 2.6 | 7.0 | 1.8 | 6.1 | 0.7 | 0.3 |
| 19 | 2037 | Ed | 215.3 | 78 | — | 0.7543 | 0.8335 | 60.4 | 12.9 | 12.7 | 1.7 | 6.6 | 1.7 | 2.4 | 1.2 | 0.4 |
| 48 | 2871 | Ed | 276.8 | 84 | 90.8 | 0.7396 | 0.8264 | 69.2 | 13.0 | 9.4 | 1.2 | 4.2 | 1.4 | 0.6 | — | — |
| S-80 | 3061 | Ed | 344 | 100 | 160.4 | 0.6870 | 0.8147 | 66.9 | 14.2 | 9.3 | 1.3 | 4.4 | 1.1 | 2.0 | 0.2 | 0.7 |

* Samples were analysed by the Jiangsu Institute of Petroleum Exploration and Development.

**Table 2. Composition of the gases released from a single in-plate oil sample from well CH-22 through stepwise Pressure decrease**

| Pressure ($1\times10^5$ Pa) | Gas composition( $V/V$,% ) | | | | | | | | |
|---|---|---|---|---|---|---|---|---|---|
| | $C_1$ | $C_2$ | $C_3$ | $iC_4$ | $nC_4$ | $iC_5$ | $nC_5$ | $CO_2$ | $N_2$ |
| 120.8 | 78.7 | 9.5 | 5.0 | 0.5 | 1.3 | 0.2 | 0.4 | 1.0 | 3.3 |
| 100 | 78.2 | 10.1 | 5.2 | 0.6 | 1.5 | 0.3 | 0.4 | 1.0 | 2.7 |
| 76.7 | 75.7 | 12.0 | 6.3 | 0.6 | 1.9 | 0.4 | 0.5 | 0.9 | 1.7 |
| 52.0 | 71.7 | 14.1 | 7.9 | 0.8 | 2.1 | 0.4 | 0.6 | 0.9 | 1.2 |
| 36.8 | 64.4 | 18.0 | 10.6 | 1.0 | 3.0 | 0.5 | 0.7 | 1.0 | 0.4 |
| 24.0 | 52.6 | 22.4 | 15.3 | 1.5 | 4.7 | 0.8 | 1.2 | 1.0 | — |
| 14.0 | 36.7 | 26.4 | 22.6 | 2.5 | 6.6 | 1.3 | 1.8 | 1.0 | — |
| 7.2 | 2.3 | 17.0 | 38.0 | 7.1 | 21.0 | 4.9 | 6.4 | 0.7 | — |

**Table 3. Depth of the oil-producing horizons and properties and composition of the crude Oil***

| Well depth (m) | GC main peak | $\frac{C_{28+29}}{C_{21+22}}$ | Light hydrocarbon (%) | Saturated hydrocarbon (%) | Aromatic hydrocarbon (%) | Resins (%) | Asphaltenes (%) |
|---|---|---|---|---|---|---|---|
| 1338 | 22 | 0.59 | 10.7 | 62.3 | 12.4 | 13.5 | 1.1 |
| 1390 | 22 | 0.48 | 18.7 | 61.4 | 8.1 | 9.0 | 2.8 |
| 1500 | 17 | 0.80 | 28.5 | 37.2 | 10.7 | 21.9 | 1.7 |
| 1816 | 22 | 0.66 | 8.2 | 70.4 | 10.4 | 9.8 | 1.2 |
| 1819 | 17 | 0.68 | 34.3 | 46.9 | 8.5 | 7.0 | 3.5 |
| 2005 | 17 | 0.46 | 21.2 | 46.0 | 11.7 | 15.2 | 5.8 |
| 2036 | 22 | 0.38 | 10.7 | 63.4 | 6.9 | 17.8 | 1.3 |
| 2064 | 22 | 0.43 | 13.5 | 65.5 | 8.0 | 8.4 | 4.7 |
| 2254 | 22 | 0.47 | 11.4 | 72.0 | 9.2 | 6.4 | 1.0 |
| 2308 | 22 | 0.59 | 35.9 | 50.5 | 7.6 | 4.8 | 1.3 |
| 2432 | 23 | 0.72 | 12.2 | 67.8 | 7.3 | 4.7 | 5.7 |
| 2481 | 23 | 0.69 | 20.1 | 55.7 | 8.9 | 11.2 | 4.0 |
| 2705 | 15 | 0.44 | 15.8 | 69.9 | 6.5 | 5.3 | 1.9 |
| 2859 | 22 | 0.63 | 19.5 | 64.6 | 6.6 | 6.0 | 3.3 |
| 3047 | 22 | 0.35 | 16.6 | 67.4 | 4.7 | 10.5 | 0.8 |
| 3062 | 22 | 0.68 | 1639 | 62.8 | 8.1 | 8.8 | 3.5 |
| 3102 | 22 | 0.76 | | | | | |
| Correlation coefficient | | +0.319 | −0.058 | +0.349 | −0.681 | −0.396 | +0.128 |

* From He Zhigao(1981).

## 2.2 Representative samples of petroleum hydrocarbon gases

Based on the data mentioned above and in the light of the principle of selecting representative samples, the oil-soluble gas of in-place oil from Well S-80 is selected for the representative of petroleum hydrocarbon gases, and its composition is presented in Table 1. The sample is characterized by the greatest sampling depth, the highest formation pressure, $C_2 > C_3$, a very low nitrogen content (only 0.2%) and a low carbon dioxide content (less than 1%). Therefore, it can be considered that the gas has been neither mixed with air nor subjected to secondary oxidation and action of microorganisms. The sample has a relatively high methane content, which was not caused by thermal degradation of the crude oil. Because of the high saturation pressure and high dissolubility of gas in oil, the specific gravity of the subsurface oil is rather low, which indicates that this gas sample has maintained the characteristics of the original composition of the gases of the crude oil to the greatest extent. Thus the gas dissolved in this crude oil is the best representative of the petroleum hydrocarbon gases, and its composition is undoubtedly the closest to that of the gases in their original formative state as various interferences and epigenetic changes have been eliminated.

## 2.3 Composition of fatty acids in fermentation liquor for marsh gas generation

The composition of fatty acids in fermentation liquor for artificial marsh gas generation varies with different fermentation media and conditions. In order to compare with the Northern Jiangsu crude oil under study, animal excrements were selected for the raw material for the liquor; the composition of its fatty acids are shown in Table 4.

## 2.4 Correlativity

Through a study of the rules of isomers for the formation of hydrocarbon by decarboxylation of fatty acids, it has been found that there is an intimate correlation between the fatty acids treated according to the rules of isomers and oil-soluble gaseous hydrocarbons which is mainly expressed in the consistency between the forms of the curves of their content changes and a high correlation coefficient between the two sets of data (Fig. 1.).

**Table 4. Composition of fatty acids in fermentation liquor of marsh gas** *

| Name of fatty acids | Acetic acid | Propio-ric a. | Isobuty-ric a. | $n$-buty-ric a. | Isopen-tanoic a. | $n$-pent-anoic a. | Total butyric a. | 1/2isopentantic a. + $n$-pentanoic a. |
|---|---|---|---|---|---|---|---|---|
| Content wt. % | 63.6 | 18 | 4.2 | 9.8 | 2.8 | 1.4 | 14 | 2.8 |

* Data from D. L. Klass.

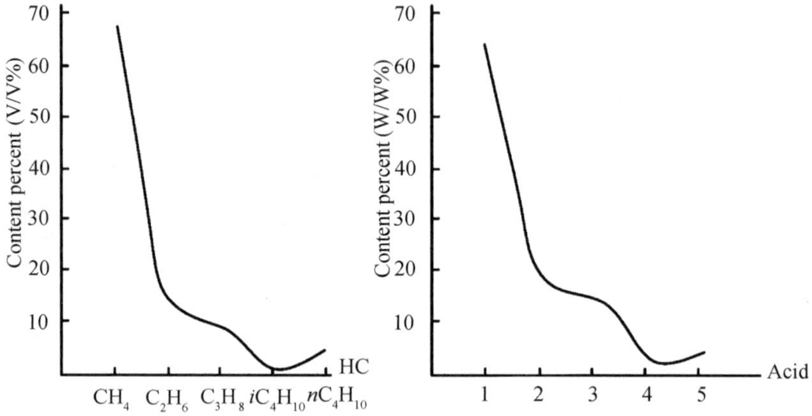


1. $CH_3COOH$, 2. $CH_3CH_2COOH$, 3. iso-butyric and $n$-butyric,

4. 1/2iso-pentenoic, 5. $n$-pentenoic and 1/2 iso-pentenoic

**Fig. 1. Curve of the compositional distribution of oil-soluble gaseous hydrocarbons (left) and normal fatty acids (right)**

(1) **Rules of isomers** According to Markovnikov's theory and general principles, there exist specific rules for the formation of hydrocarbons by decarboxylation of fatty acids. These rules imply that, though the structures of $n$-butyric acid and isobutyric acid are quite different, the same kind of hydrocarbon—propioric acid—is formed after their decarboxylation and that pentanoic acid has three kinds of isomers, of which 2-methyl-butyric acid and $n$-pentanoic acid form $n$-butane after decarboxylation and 3-methyl-butyric acid forms isobutane after decarboxylation. This decarboxylation model can be explained not only theoretically but also compared quantitatively with the cases having such relations. Their reactions are as follows:

a. $n$-butyric acid + isobutyric acid ⟶ protane

$$CH_3 \cdot CH_2 \cdot CH_2 \cdot COOH \longrightarrow C_3H_8 + CO_2$$

$$\begin{array}{l} CH_3 \cdot \underset{\displaystyle \underset{\displaystyle CH_3}{\bullet}}{CH} \cdot COOH \longrightarrow C_3H_8 + CO_2 \end{array}$$

b. 2-methy-butyric acid (1/2 isopentanoic acid) + *n*-pentanoic acid ⟶ *n*-butane

$$CH_3 \cdot CH_2 \cdot \underset{\underset{CH_3}{\bullet}}{CH} \cdot COOH \longrightarrow nC_4H_{10} + CO_2$$

c. 3-methyl-butyric acid (1/2 isopentanoic acid) ⟶ isobutane

$$CH_3 \cdot \underset{\underset{CH_3}{\bullet}}{CH} \cdot CH_2 \cdot COOH \longrightarrow CH_3 \cdot \underset{\underset{CH_4}{\bullet}}{CH} \cdot CH_3 + CO_2$$

(2) **Very similar compositional curves** Following the rules mentioned above, fatty acids are decarboxylated to form corresponding hydrocarbons, and the percentages of various fatty acids are calculated by normalizing these fatty acids. The curves of their percentages and those of oil-soluble gaseous hydrocarbons of S-80 in Table 1 are constructed separately (Fig. 1). As shown in Fig. 1, the left curve shows the compositional distribution of oil-soluble gaseous hydrocarbons, while the right one shows that of fatty acids.

(3) **High correlation coefficient** Through processing by use of mathematical statistics, the correlation coefficient of the percentage composition (vol./vol.) of the oil-soluble gaseous hydrocarbons of S-80 in Table 1 versus the percentage composition (wt./wt.) of the fatty acids in Table 4 is rather high, being up to 0.9935. This indicates that there is an intimate relation between the two sets of data and a very close similarity between the two kinds of processes.

## 3 Discussion

### 3.1 Thermodynamic analysis on fatty acid decarboxylation

Acetic acid is decarboxylated into methane; the thermodynamics of the reaction is discussed below.

$$CH_3 \cdot COOH(l) \longrightarrow CH_4(g) + CO_2(g)$$

| | $CH_3 \cdot COOH(l)$ | $CH_4(g)$ | $CO_2(g)$ |
|---|---|---|---|
| $\Delta H_f^0$: | −487.0(−116.4) | −74.8(−17.89) | −393.5(−94.05) |
| $S^0$: | 159.8 (38.2) | 186.3 (44.5) | 213.6 (51.06) |
| $\Delta G_f^0$: | −392.5(−93.8) | −50.8(−12.14) | −394.4(−94.26) |

Thus on calculations, $\Delta H = -18.66\ kJ \cdot mol^{-1}$ ($-4.461\ kcal \cdot mol^{-1}$), $\Delta S = 240.0\ J \cdot deg^{-1} \cdot mol^{-1}$ (or $57.36\ cal \cdot deg^{-1} \cdot mol^{-1}$), and $\Delta G =$

$-52.7\ kJ \cdot mol^{-1}$ (or $-12.6\ kcal \cdot mol^{-1}$). From above, it can be seen that the acetic acid decarboxylation is an exothermic, entropy-increase and free energy-decrease reaction under standard thermodynamic conditions.

According to the CRG rule, $\Delta G^0$ values (values of free energy changes) for the decarboxylation of propioric acid and higher-molecular-weight volatile fatty acids are all about $-41.84\ kJ \cdot mol^{-1}$, and all fall in the range of the energy level of the biophysiological effect ($33.5 - 50.2\ kJ \cdot mol^{-1}$)[3].

## 3.2 Geochemical analysis on fatty acids and their decarboxylation

As is well known to all, fatty acids are widely distributed in crude oils and ancient and recent sediments and show a preference for even-carbon-numbered molecules.

K. A. Kvenvolden et al. (1965, 1969)[4,5] suggested that the predominance of even-carbon-numbered *n*-fatty acids is smaller in ancient sediments than in modern sediments and that petroleum *n*-alkanes are derived from decarboxylation of *n*-fatty acids but are one carbon number less than their corresponding fatty acids because of decarboxylation. That is to say, the predominance of even-carbon-numbered fatty acids in modern sediments becomes small in source rocks after diagenesis. By inference, the odd carbon predominance of *n*-alkanes should be increased.

It is evident, however, that there exists a contradiction between the changes of even-odd predominance of fatty acids and those of alkanes: the odd carbon predominance of *n*-alkanes in rocks does not increase as stated above, but decreases. If the odd carbon predominance of *n*-alkanes in ancient rocks decreased or the even carbon predominance increased, then the even carbon predominance of fatty acids in rocks should not decrease, but on the contrary the even/odd ratio (or referred as to the acid distribution ratio) should be higher in ancient sediments than in modern sediments owing to a decrease in odd carbon number, i. e., the even/odd ratio should become higher when its denominator becomes smaller. This is obviously inconsistent with the fact.

A lot of geochemical data indicate that both of the odd carbon predominance of *n*-alkanes and the even carbon predominance of fatty acids decrease with increasing burial depth of the rocks. Therefore petroleum *n*-alkanes in rocks are by no means formed by decarboxylation-thermal degradation of normal fatty acids that are one carbon number more than *n*-alkanes. In other words, the rules of isomers discussed in the preceding section of this paper do not exist in rocks, nor do the decarboxylation of fatty acids and the quantitative relation.

Theoretically, subsurface temperatures may lead to fatty acid decarboxylation, but can not cause the marked difference in odd/even carbon numbers. From the practical data, fatty acids, regardless of odd or even carbon numbers, can exist in rather deep wells and so do amino acids, both of which are quite stable compounds. Besides there exist predominances of even-carbon-numbered $n$-alkanes in ancient rocks such as carbonate rocks and evaporites. It is worthy to note that there are different carbon number predominances in different carbon number regions for the same sample. Such a contradictory phenomenon is difficult to explain by the thermal degradation theory for oil generation.

The author considers that gaseous hydrocarbons are derived from light-molecular-weight fatty acids, which have a high solubility in water and are hardly adsorbed by solid sediments (organic inorganic matter). In fact fermentation liquor for artificial marsh gas can fully contact the fermentation materials, so the materials have been saturated with fatty acids due to adsorption. Thus the fatty acid decarboxylation producing large quantities of petroleum hydrocarbon gases can only proceed in water. From the above-mentioned thermodynamic analysis and characteristics of bioactivity, decarboxylation reaction is very likely to be accomplished by microorganisms.

Table 5 shows the compositional variation of pentanoic acid isomers with different time durations and at different portions (upper, middle and lower) of a fermentation device for artificial marsh gas—shunt reactor.

**Table 5. Relationship between the composition of pentanoic acid in the fermentation liquor and the fermentation duration and positions of the fermentation device***

| Name | $n$-pentanoic acid (mg/L) | | | Isopentanoic acid (mg/L) | | | $\frac{n\text{-pentan. a.}}{\text{isopentan. a.}}$ | | | $\frac{n\text{-pentan. a.} + 1/2\text{isopenta. a.}}{1/2\text{isopenta. a.}}$ | | |
|---|---|---|---|---|---|---|---|---|---|---|---|---|
| Time \ Portion | Lower | Mid. | Upper | Lower | Mid. | Upper | Lower | Mid. | Upper | Lower | Mid. | Upper |
| 2 h | 157 | 110 | 133 | 49 | 45.1 | 55 | 3.2 | 2.4 | 2.4 | 7.4 | 4.8 | 5.8 |
| 7 h | 99.3 | 94.1 | 123 | 41 | 47 | 43.3 | 2.4 | 2.0 | 2.8 | 5.8 | 5.0 | 6.7 |
| 15 h | 40.2 | 54.2 | 61 | 50 | 45 | — | 0.8 | 1.2 | | 2.6 | 3.4 | |

* Fermentation conditions: reactor ∅ = 6.2 cm, reaction temperature 36 ℃ - 39 ℃. Experiments performed by the Chengdu Institute, Academia Sinica.

As is shown in Table 5, isomers of fatty acids increase and *n*-fatty acids and the *n*-pentanoic/isopentanoic ratio decreases with an increase of fermentation time. Take the middle portion for example, the *n*-pentanoic acid/isopentanoic acid ratio decreases from 2.44 after 2 h to 2.00 and 1.20 after 7 h and 15 h respectively, and that is also the case with the ratio of (*n*-pentanoic acid + 1/2isopentanoic acid)/(1/2 isopentanoic acid). Such ratio changes show that the formation of gaseous hydrocarbons by decarboxylation calls for a *n*-butane/isobutane ratio of about 1.5 and more than 15 h of fermentation. The fatty acids generated within this interval possess the features of forming petroleum hydrocarbon gases. Therefore the experimental data tell us that the stage of forming petroleum gases by decarboxylation of fatty acids is later than the stage of fermentation for forming artificial marsh gases.

Comparison between the data of stable carbon and oxygen isotopes of the methane and carbon dioxides in artificial marsh gases and the fermentation time is shown in Table 6.

As shown in Table 6, such heavy gaseous hydrocarbons as ethane have not yet been seen even after 19 days of fermentation, and the carbon isotopes have not been fractionated into those of petroleum gases. Therefore, the formation of petroleum gases is not a biochemical reaction of artificial marsh gases but another specific biochemical reaction taking place after marsh gas generation.

**Table 6. Comparison between the isotope data of methene gas and natural gas**

| Fermentation time (d) | $CH_4$ $\delta^{13}C$‰ (PDB) | $CO_2$ $\delta^{13}C$‰ (pDB) | $CO_2$ $\delta^{18}O$‰ (PDB) | $\frac{CO_2}{CH_4}$ ($V/V$) | $C_2^+$ % ($V/V$) |
|---|---|---|---|---|---|
| 2 | -53.48 | -13.47 | -4.88 | 0.56 | <0.001 |
| 5 | -54.69 | -10.45 | -4.85 | 0.70 | <0.001 |
| 7 | -53.95 | -8.51 | -4.79 | 0.87 | <0.001 |
| 9 | -53.33 | -7.85 | -4.63 | 1.14 | <0.001 |
| 13 | -52.93 | -7.50 | -4.69 | 1.19 | <0.001 |
| 15 | -52.75 | -8.20 | -4.88 | 1.22 | <0.001 |
| 17 | -52.13 | -8.12 | -4.89 | 1.17 | <0.001 |
| 19 | -48.01 | -6.77 | -4.51 | 0.93 | <0.001 |
| Jiangsu natural gas | -30.5 | -5.14 | | | |

## 4 Conclusions

Being influenced by many factors, the composition of petroleum hydrocarbon gases are very complex and highly variable. After the elimination of the effects of the factors such as pressure and biodegradation, the oil-soluble gaseous hydrocarbons in high-pressure in-place oil are hydrocarbon gases really representing the original composition of petroleum gases. There is a strong correlation between the composition of this gas and that of the volatile fatty acids in marsh gas fermentation liquor processed according to a given rule. Their linear correlative coefficient is 0.9935 and their content curves are very similar. The theory of the isomer rules—fatty acids are decarboxylated into hydrocarbons—has come from practice, and, in turn, has been checked by practice.

In the light of the fact that the even-carbon preference of fatty acids and the odd carbon preference of $n$-alkanes are smaller in ancient rocks than in modern sediments, it is proposed that there is no isomer rule of fatty acid decarboxylation during the transformation from modern sediments into source rocks, i. e. , there occurs no decarboxylation of even-carbon-numbered normal fatty acids into odd-carbon-numbered $n$-alkanes which are one carbon less than the acids. For, according to the isomer rules, a decrease of the even-carbon preference of fatty acids is bound to lead to an increase of odd-carbon-numbered. But in fact, both of even-carbon-numbered normal fatty acids and even-carbon-numbered $n$-alkanes decrease with an increase of burial depth (except for evaporites, etc.). Based on these premises, the reaction of decarboxylation of fatty acids into hydrocarbons according to the isomer rules is impossible to be that of thermal, decomposition.

The decarboxylation of light-molecular-weight fatty acids must proceed mainly in water, media, because their solubility in water is high and they are scarcely adsorbed by sediments.

Thermodynamic analysis indicates that the free energy change $\Delta G^0$ of the decarboxylation reation is $41.87 - 50.2$ kJ · $mol^{-1}$, well within the range of the energy levels of biophysiological changes.

Thus decarboxylation for forming petroleum hydrocarbon gases should be a biological reaction: such a biological reaction, however, is not an ordinary biochemical one taking place in the fermentation for producing artificial marsh gases, in the light of the tendencies of changes of fatty acid isomers in the marsh gas fermentation liquor and methane carbon isotopes in marsh gases as well as the contents of ethanes and heavier hydrocarbons in marsh gases, it can be judged that biochemical reactions for producing petroleum hydrocarbon gases are special biological reactions proceeding after the fermentation for marsh gas generation.

Chinese manuscript received May 1985
accepted March 1986

## References

[1] Seifert WK. Carboxylic acid in petroleum and sediments. Fortschr. Chem Org Naturst, 1974, 32:2 -3,38 -39.

[2] Califert Y. Evolution Expérimentale d'huiles brutes et de fractions d'huiles brutes sous l'influence de la température de la pression et de minéraux [M] // Advances in Organic Chemistry. Pergamon Press, 1968.

[3] Bryant MP. Microbial methane production—theoretical aspects [J]. J Anim Sci, 1979,48(1): 193 -201.

[4] Kvenvolden KA. Evidence for transformation of normal fatty acid in sediments[M] // Advances in Organic Geochemistry. Pergamon Press, 1966.

[5] Kvenvolden KA. Geochemical exploration, "United States Patents" 3, 1969, 480, 396,11.

本文原载于 *Acta Geologica Sinica* (English Edition) No. 60, 1986.

# 5. 论石油形成体系的化学基础——化学反应方程集合——油田形成的最大奥秘

# On the chemical basis of the formation system of petroleum—Collection of chemical reaction equations—The biggest mystery in oilfield formation

何志高

（中国石化勘探开发研究院无锡石油地质研究所）

**摘要：**本文在化学理论基础上论证石油的生成和揭示油田形成的最大奥秘。

化学基础是由4个化学反应式组成，并且赋予它们一种数学观念——“集合”概念，称为“石油成因化学方程集合”，并以图展现该集合。集合中全部化学反应都是为了一个生成石油烃类的目标化学反应，有一完整的化学链，是缺一不可的整体。而且这个体系是一种高效化学反应体系，这个复杂的石油形成体系，其深刻内涵一目了然。

对化学反应的研究，采用基本的热力学方法和质量做用定律判别反应进行方向的科学合理性和必然性。由这些基本方法和理论可以确定，碳酸盐对石油生成反应具有极其重要的作用。

碳酸盐与生烃化学反应中的反应产物二氧化碳发生化学反应，生成可溶于水的重碳酸盐，从而有效、有力地保证了生烃反应长期而稳定地进行。碳酸盐的来源是丰富的：来源之一是碎屑沉积碳酸盐，二是重碳酸盐的分解反应所产生的碳酸盐沉淀。而重碳酸盐来源于沉积物陆源区的碳酸盐风化溶蚀后进入沉积盆地水体。

重碳酸盐的分解是普遍发生在地质体和自然界中的化学反应。它的产物还有另外多个既特殊又极为重要的作用：一是二氧化碳作为高效光合作用的碳源，产生了能生成石油烃类的能转化为脂肪酸的糖类以及组成约占一半石油成分的胶质和沥青质的原始有机物质；二是从水体中析出的碱土金属元素等具有纳米级物质特效，为水体生物繁衍提供丰富的物质来源，例如镁对叶绿素无疑具有巨大作用。

根据石油成因体系化学理论分析，很自然地揭示出油田形成的尚不为人知的最大奥秘：有碳酸盐古物源参与建造的湿封闭体系是形成油田的必要条件，没有碳酸盐就没有油田。这种结论是能够经得住大量石油地质实践检验的。世界上最大的加瓦尔油田就是最好的验证。笔者创立的有机生油理论铁证生油剖面毫无疑问地是这种理论的铁证。笔者在文中还有关于中国油田的点评以及页岩气的相关评论。

本文是笔者在40多年关于石油成因研究的基础上所做的最新研究成果论述。查阅笔者历年文章，可以省去大量参考文献的罗列，有益于追索证据以及避免重复。

**关键词：**石油生成化学基础；化学反应方程集合；高效化学反应体系；油田形成最大奥秘；碳酸盐古物源；湿封闭体系

**Abstract**: This paper demonstrates the oil origin and to reveal mystery the oil field forming on the based of the chemical theory.

The chemical foundation is determined by the 4 chemical equations and gave them a mathematical concept—Collection, a collection of the chemical equations for the petroleum origin, and to show in drawing. All the chemical reactions are for a generation of petroleum hydrocarbons, a complete chemical chain, is an indispensable integral. This is a higher activity chemical system. Stick out a mile deep connotation.

Study on the chemical equation, using basic thermodynamic method and the Law of Mass Action, to judge the scientific rationality and inevitability of those reactions. After the study identified: the carbonate plays a very important role in petroleum formation reaction. The carbonate reacts with carbon dioxide that is reaction of the forming oil products to form soluble bicarbonate. Therefore, the source responses to long-term and stable.

The carbonate source is rich. It is clastic sedimentary carbonate, the precipitation of carbonate which decomposition of the bicarbonate; The bicarbonate that it is one into the sedimentary basin water is source by weathering denudation of the carbonate derived from terrigenous sediments area.

The decomposition of bicarbonate is the chemical reaction, which is very easy and widespread. Its products have yet 3 special and extremely important role: one of them is the carbon dioxide as the carbon source for photosynthesis; other one of them is this kind of special causes can produce alkaline earth metal with nanometer effects, such is the role of magnesium chlorophyll.

Analysis by the origin of petroleum system of chemical theory, naturally reveal the greatest mystery of oil formation: a large number of carbonate can only originate from the basin surrounding land, provid sediment sources that is paleogene provenance of the carbonate(ancient source material); without this carbonate is no oil field.

This conclusion is able to withstand large amounts of petroleum geology of the actual inspection. Ghawar field and cold source profiles are very good verification.

This paper also reviews about shale gas.

20 世纪 80 年代,湿封闭体系综合效应成油论[1-3]突破传统生油理论模式,开创了石油成因理论崭新时代,之后,完成了一系列的相关理论建树,如泥质岩不是生油岩[4]、煤不成油的理论基础[5]、论干酪根热降解成油理论弊端[6]、陆相石油与生油理论[7]、关于碳酸盐岩的生油研究[8]、关于石油成因理论发展历史若干问题的思考与评价[9]等十多个相关专题。同时,应用新理论讨论台湾油气勘查选区与油气性质评价和台西盆地油气组成与性质差异的成因分析[10,11];在理论上,应用化学热力学的基础资料,提出天然气组成特点及其成因机制分析[12],发表《煤成气量的热化学理论计算方法与从热力学理论讨论烃类的演化》[13,14]等论文。2011 年,笔者发表了《"中国加瓦尔胚胎"与有机成油理论铁证生油剖面及一种新命名的岩石——"油泥灰岩"的命名申述》[15]一文,笔者认为,这篇文章把湿封闭体系成油理论从原来的以实验室为主、以推理为主、以微观为主上升到了有一条地质体中厚度近 1 km、全岩心、宏观与微观兼备的综合剖面,铁证般地呈现出有机成油理论的科学依据和基础,这是一条可以重复展现的、科学的、任何人可以再钻井再取岩心再研究再得出相同原理结论的具有标准剖面性质的地质体。因此，有理由说，有机成油理论不应该再称为“假说”，假说的历史应该结束。“湿封闭体系成油理论”是科学的理论。

本文是在上述40多年研究的基础上的最新研究成果论述，是从化学理论基础上论证石油的形成，揭示油田形成的奥秘，这无疑会对全球石油勘探、开发极具新的指导意义。什么盆地有石油和无石油或者油多油少都会为石油地质学家提供新的判断研究理论依据。

## 1 石油形成体系化学基础

### 1.1 关于石油烃类生成的代表化学反应式

根据研究[1-3]确定石油烃类生成的代表化学反应式为：

$$RCOOH \xrightarrow[\text{微生物}]{\text{水介质}} RH + CO_2 \tag{1}$$

(1) 式中，RCOOH代表各种生成石油烃类的脂肪酸类；RH代表各种石油烃类；$CO_2$代表二氧化碳。反应式(1)是微生物的生物化学作用，介质为湿封闭体系。

众所周知，对化学反应进行研究的基本方法是应用化学热力学和化学反应动力学。

笔者早年已从理论和实验室实验方法确定，(1)式的化学反应热力学性质是在标准状态下，焓变化值为负值($\Delta H<0$)，放热，熵增加($\Delta S>0$)，自由焓降低($\Delta G<0$)，这3个基本参数判别是能自发进行的化学反应。微生物的实验结论是[5]，在25 ℃、1个气压条件下，生成反应是自动快速可以多次重复完成的反应。

关于(1)式反应动力学性质的问题，虽然一般而言，地质体中的化学变化的时间相对漫长，时间是一个相对较大的参数，相关速率函数的定量分析在理论上就将失去意义。然而，由于这个问题的重要性，我们还是需要而且有可能予以研究讨论。从自由焓和化学反应的平衡常数的关系，即$\Delta G$值的正负符号和绝对值的大小可以准确定性判定与半定量地确定反应动力学问题。虽然这也是一个涉及更加复杂的封闭体系与变组成体系的问题。我们将在“讨论”中再论述。

笔者认为，化学反应中的质量作用定律更能使我们认识(1)式的特性，进而认识生成石油烃类的反应的独特与神奇。

依据质量作用定律的基本原理，在非气态分子碰撞与忽略地质体中时间因素决定的反应速率的体系，要使(1)式反应长期保持自动进行状态，在不涉及论证反应物，即(1)式左边的 RCOOH 的前提下，生成物 RH 增加，另一个反应产物二氧化碳($CO_2$)必须减少，或者 RH 和 $CO_2$ 二者都减少。

非常有趣的问题是石油烃类 RH，几乎所有烃类(包括大多数石油非烃类)都是不溶于水的。这就是说，反应产物 RH 自动离开总的反应体系——液相生物化学反应体系，即反应产物进入另一个不相容的体系。

按照通常的一般的化学反应，随着反应的进行，反应产物浓度(压力)增加，必然会影响该反应的平衡常数及该反应的热力学参数等。然而，石油烃类的特殊性质，成就和创造了石油烃类本身可以大量生产的一种优越条件。无论气相和液相烃类生成都对生成化学反应影响很小，甚至可以认为在一个特定储聚空间可以忽略不计。油田开发中的高压事实上是沉积物深埋增加所致，并非石油烃类生成时期体系的压力状态。这是一个在石油烃类生成的化学机制上的具有重要理论和实际价值的新理念。

另一个更加有趣、更加重要的问题是，(1)式反应产物中另一个反应产物二氧化碳的性质、行为与产生的效果。

在化学反应平衡理论上，$CO_2$ 与 RH 一样，都是一个化学反应式中的反应产物，它也会影响生成石油烃类的反应式(1)的进行。然而，$CO_2$ 与 RH 在性质上又不一样，它在水中有相当大的溶解度。然而，为了反应式(1)的自动向右方进行，$CO_2$ 必须去除，否则，随着生烃反应的进行，$CO_2$ 增加，体系压力增大，$CO_2$ 在液相中的溶解度增大，必然会阻碍反应式(1)的进行，甚至终止生烃。

### 1.2 关于消除阻碍石油烃类生成的化学反应式——碳酸盐的溶蚀溶解作用

根据研究与大量的地质实际情况，在湿封闭体系的地质环境中，能减少或除去二氧化碳，确保反应式(1)式顺利长期进行的唯一途径是碳酸盐及其化学反应。以碳酸钙代表碳酸盐，它与二氧化碳反应，以除去二氧化碳，生成可溶性的重碳酸盐，例如碳酸氢钙，其化学反应式如下：

$$CaCO_3 + CO_2 + H_2O \xrightarrow{\text{水介质}} Ca(HCO_3)_2 \tag{2}$$

对比(2)式与(1)式可以看到，(2)式的反应物正是(1)式的生成物。由上述质量作用定律可以判定，(2)式确保(1)式的进行，(2)式的 $CaCO_3$ 越多，越有利于(1)式向生成石油烃类的方向进行；同理反之，(1)式产生的 $CO_2$ 越多，也必然会使(2)式向右方向进行。没有(2)式就没有(1)式，没有大量的碳酸盐就没有大量石油烃类的生成。

### 1.3 关于第3个石油形成体系的重要化学反应式——重碳酸盐在水溶液中的分解反应

笔者认为，下述化学反应式(3)是对石油生成和油田形成有三大贡献的化学基础：

$$Ca(HCO_3)_2 \xrightarrow{\text{水介质}} CaCO_3\downarrow + CO_2 + H_2O \qquad (3)$$

反应式(3)看似是上述反应式(2)的逆反应，但是，它们不是一般意义上的化学反应的化学平衡(正逆)反应表达式上的关系，而是在地质体不同领域、不同时间发生的各具特征作用的化学反应。该反应的三大贡献之一是，沉淀 $CaCO_3$ 在水介质条件下与反应式(2)中的 $CO_2$ 完成反应式(2)，其贡献可谓大也；二是具有纳米级特性的碳酸盐在沉积盆地水体中为水生生物提供了宝贵的钙、镁等物质源之需；三是在沉积水体中 $CO_2$ 为光合作用反应式(4)提供碳源，大量地繁殖了水生生物，从而为生成石油烃类和非烃类提供反应式(1)所需的脂肪酸之源。

### 1.4 关于光合作用与成油关系的化学问题

传统的有机生油理论仅仅提到水生生物与陆源有机物质。笔者认为，生油有机物质来源主要是上述反应式(3)提供的 $CO_2$ 所进行的光合作用而获得生物碳源，而且，这种光合作用还包括进入水体中的碳酸盐和重碳酸盐提供的钙、镁和其他微量元素，特别是叶绿素-镁，既是植物的组成部分，又是光合作用的催化剂等。碳酸盐—光合作用—生油有机物质源，这种关系非常重要、非常清楚。本文将这种光合作用稍作注释，仍然用通常的光合作用化学式(4)代表如下：

$$CO_{2[Ca(HCO_3)_2]} + H_2O \xrightarrow[\text{光合作用}]{\text{水介质}} O_2 + \frac{1}{6}C_6H_{12}O_{6[RCOOH]} \qquad (4)$$

笔者对式(4)中 $CO_2$ 右下方加注 $Ca(HCO_3)_2$，其意义是突出这种水体中的光合作用的 $CO_2$ 来源是限定的 $Ca(HCO_3)_2$；$C_6H_{12}O_6$ 的右下方加注 RCOOH 的用意也是突出这种糖类是限定的 RCOOH 之源。本反应的反应产物氧无疑也有新生态的特殊性质。

### 1.5 关于化学反应方程集合

从上述可见，4个化学反应式相互之间紧密联系，并且具有石油烃类生成全过程的综合整体性质。为了说明、理解和强调这种特殊的关系和性质，笔者认为，引进一个数学中的基本概念——集合(表达具有一种特定性质的事物的总体)是很有意义的。这样，有利于深刻揭示石油成因理论。

我们取 $A$ 代表具有成油化学基础特性总体的集，4个单独的化学反应式分别用符号 $a_1, a_2, a_3, a_4$ 代表组成该集合的元素，则该集合可用图表示如下：

$$(a_1,a_2,a_3,a_4)\in A—\left[\begin{array}{l} a_1\ \cdots RCOOH \xrightarrow[\text{微生物}]{\text{水介质}} RH + CO_2 \\ a_2\ \cdots CaCO_3 + CO_2 + H_2O \xrightarrow{\text{水(空气)介质}} Ca(HCO_3)_2 \\ a_3\ \cdots Ca(HCO_3)_2 \xrightarrow{\text{水介质}} CaCO_3\downarrow + CO_2 + H_2O \\ a_4\ \cdots CO_{2[Ca(HCO_3)_2]} + H_2O \xrightarrow[\text{光合作用}]{\text{水介质}} O_2 + \frac{1}{6}C_6H_{12}O_{6[RCOOH]} \end{array}\right]$$

**图 1　石油成因化学反应方程集示意图**

“集”的概念是有物质基础的，这个基础就是所有化学反应都是在水介质环境下，所有化学反应都是为了一个生成石油烃类的目标，是一个完整的化学链，是缺一不可的总体。一个复杂的石油形成体系，从图 1 可以一目了然其深刻内涵。

## 2　油田形成的最大奥秘

根据上述所确定的生成石油烃类的第一个化学反应式，我们首先以提问的方式来揭示油田的最大奥秘。然后再以 4 个反应式之间的紧密联系以及化学反应方程集示意图来论证这个最大奥秘的合理性、科学性。

问题 1：为了使生成石油的反应长期顺利进行，二氧化碳必须从体系中去除（质量作用定律），这是肯定的。那么，怎么去除二氧化碳呢？

答案：为体系提供碳酸盐，如反应式(2)中的 $CaCO_3$。

问题 2：体系中的碳酸盐从何而来？

答案：碳酸盐来源有两个：一是沉积碎屑碳酸盐，二是重碳酸盐的分解反应产生并沉淀沉积，如反应式(3)中的 $CaCO_3$。

问题 3：碎屑碳酸盐和重碳酸盐从何而来？

答案：碎屑碳酸盐来自古物源；而重碳酸盐则来自古物源碳酸盐风化和水溶蚀而随同碎屑物一起进入沉积体系，如反应式(2)。

问题 4：反应式(1)中的脂肪酸从何而来？

答案：体系中的脂肪酸来源有两个：一是陆源有机物质生物化学等作用产物；二是重碳酸盐分解产物中的二氧化碳在水体中的光合作用所获得有机物的生物化学作用产物。

上述 4 个问题可以直接追溯到沉积物的最早源头，地质学称之为“古物源”，即来自沉积盆地古地理时期给盆地内提供物质的周边陆地（大山等）。能提供碳酸盐的古物源就是盆地形成油田的关键。

大家知道,海相碳酸盐油田,特别是世界上最大的加瓦尔油田,美国地质学家确立的生油岩是富含有机质的泥灰岩,储油岩是碎屑灰岩,从生到储都是碳酸盐。笔者所确立的有机生油铁证生油剖面是碳酸盐构成的剖面。

上述 4 个反应式所发生的环境都是水介质环境,均为含水的体系。没有水就没有任何一个反应,就没有石油烃类的生成,这也是 4 个化学反应能够统一,能够成为一个化学反应方程集的物质基础,这就是笔者 20 年前创立的“湿封闭体系成油理论”中的“湿饱和水体系”的基础。

上述 4 个化学反应式中的每一个反应式都是具有坚实化学基础而且在地质体和自然界广泛存在的无可争辩的事实。

经过反复论证与分析,完全可以认为碳酸盐是上述 4 个反应的核心和基础,是石油烃类生成的核心和基础,它保证了反应式(1)的进行。碳酸盐的来源对一个沉积盆地来说,只能是陆源。

沉积古物源,碳酸盐或以碎屑状态或以反应式(2)的形式进入沉积盆地之中。碎屑碳酸盐和反应式(3)产生的碳酸盐吸收反应式(1)的反应产物 $CO_2$,顺利完成生成石油烃类的过程和结果。毫无疑问,要生成大量的石油,必须有大量的碳酸盐。

沉积古物源碳酸盐按反应式(3)生成 $CO_2$,在光反应和暗反应的光合作用下为水生生物繁衍提供大量营养,这巨量的糖类当然成了生成石油烃类反应式(1)的物质基础。

沉积古物源的可溶性碳酸盐,除了可溶性的重碳酸盐外,水的溶蚀风化也必然将其中的铁、镁、锰、铜、锌等水生生物所需要的微量元素带入水体系统,进而随同形成烃类的有机物质和成烃环境,最终形成石油非烃类。

昭然若揭,生成石油烃类和形成油田的最大奥秘就是:沉积盆地具有足量和适时的碳酸盐古物源参与构建的湿封闭体系是油田形成的最重要的基本条件;没有这种碳酸盐,就没有油田。

## 3 讨论

石油地质方面必须充分研究和理解地面上所看到的碳酸盐地区卡斯特地貌,笔者认为那些溶洞、洞穴、凹陷、地下河、地下湖等在碳酸盐油田的地质体中都是可以存在的。这种碳酸盐的特性在沉积成岩期形成油田的全过程都具有普遍性,而且在某些方面还会比地面更加强烈,纵向上、横向上的溶蚀作用更加向深度和广度扩展。因为,地下成油体系是一个封闭体系,$CO_2$ 的压力和产生烃类气体的压力会增加 $CO_2$ 在水中的溶解度,从而增加这种溶蚀作用的强度。大量生油的有生物溶解的液体必然极大地增强这种溶蚀作用。这是主力生油岩与储油岩一体的必然，是自生碳酸盐和陆源碎屑碳酸盐的高度混杂，是石油地质学家与岩石学家常常遇到的难题之一，也是石油地球化学家遇到的分析测试结果巨大差异的难解难题的原因之一。笔者将这种岩石定名为"油泥灰岩"。

碳酸盐的作用是多方面的，除了反应式中的 $CaCO_3$、$CO_2$ 以外，钙离子、钙元素和镁离子、镁元素等在沉积水体的水生生物繁衍中的作用也是十分巨大的。尤其在占原油成分约一半的非烃类的胶质沥青质的起源上，作用更为巨大而独特。例如叶绿素中的镁、动物血红素中的铁等都可以由碳酸盐岩石提供。

进入水体中的 $CO_2$，在造就生油有机物质上蕴含着极为深刻而广泛的作用。我们可以通过海洋及其相关生命科学了解到 $CO_2$ 在生物代谢等方面起到的巨大作用，而笔者的研究认为，成油盆地沉积水体中起着极为巨大作用的是碳酸盐矿物，即化学反应式(3)提供的 $CO_2$[16]。

在陆相油田，笔者在陕甘宁地区研究了大量的含油砂岩与非含油砂岩的物性分析测试与薄片鉴定资料，含油砂岩物性好，渗透率和孔隙度高，胶结物很少甚至完全没有碳酸盐胶结物。而非含油砂岩相反，物性很差，碳酸盐硅酸盐胶结物填满孔隙。这是几十年来千千万万个样品所确定的事实。几代人都把这种事实解释为原油对储油岩的腐蚀溶解作用。当然也有人认为储油岩沉积时就没有碳酸盐，正好这些孔隙可以接纳从生油岩中运移来的原油；还有"次生说"；等等。

笔者认为，含油砂岩体系本身就是一个生油的湿封闭体系。陆相含油气盆地都具备本文上述 4 个化学反应式的条件，所以，它能生油，能成油田。

页岩、油页岩、煤等可燃矿产的油气地球化学问题，都可以应用本文上述 4 个化学反应式组合的化学基础，研究评价开发可燃气体和能源。当今世界盛行页岩气的开发，本文提出的这些理论可能具有重要作用。一般而言，通常页岩的主要成分是泥质黏土类，孔隙度小，含水少，脱水较快，微生物的作用较弱，随同沉积的不多的有机物质所进行的生物化学作用只能是时间不长的生成少量烃类气体并以甲烷为主的最终结果。如果一种页岩含砂量大，陆源沉积物有碳酸盐(包括自生碳酸盐或者还有碎屑碳酸盐)，即有湿封闭体系，也必然会有 4 个化学反应式组合的成烃、成油的化学基础。它相对砂岩来说，生烃生油的强度和能力较低，运移能力较低，然而它的泥质成分却具有较强的吸附烃类气体的能力，如果它的保存条件好，这类页岩可能就是巨大能源资源贮集藏。

笔者多篇文章论述了煤不成油，并且推断煤肯定不成油的结论是有坚实的理论基础的。然而，煤系地层产油、产煤盆地产油是因为具备本文所论述的石油成因化学反应方程集所构成的湿封闭成油体系。而煤焦油则是无水条件下加热干馏的产物。同理，干酪根成油理论与煤焦油的形成机制是一致的。至于玉门老君庙油田等中国和外国的油田，由于“红色泥岩红色地层找不到生油岩”，只能当石油成因历史笑料了。

### 3.1　成油过程中碳酸盐的两个特殊性质对应两个重要作用

重碳酸盐的水溶液呈碱性，有利于微生物的繁衍、增殖，这是有利于石油烃类生成的。在化学上，重碳酸盐离解成碳酸氢根，与水分子离解的氢离子结合成较稳定的碳酸，从而提升水溶液中氢氧根的离子浓度，提高 pH 值。

重碳酸盐的分解反应（反应式 3）所析出 $CaCO_3$ 的沉淀，在一段时间内呈现出纳米级的特性，这也非常有利于生物生长繁衍，增加沉积有机物质的总量。

### 3.2　反应式（1）的动力学方面的问题讨论

根据热力学基本原理，任何一个反应均有关系式 $\Delta G^0 = -RT\ln K$，都可以由此关系式求得平衡常数。笔者早期求得的反应式[1]的 $\Delta G^0$为 $-52.7$ kJ/mol（$-12.6$ kcal/mol）(1)，其平衡常数即可通过下列计算求出：

$$-12600 = -1.987 \times 298.15 \times 2.303 \times \log K$$

$$K = 1.72 \times 10^9$$

由此可见，这是一个相当大的平衡常数。笔者认为，从理论上来看，绝对温度参数以及其他相关参数都不可能使平衡常数出现负值，在很低温度范围都是能自发进行的反应；这种计算结果与实验室的实际和地质体中的实际是完全一致的，理论和实际完全相符。反应式（1）可以认为实际上的不可逆反应，接近于理论上定义的平衡常数 $K$ 趋于无穷大；也是实际上属于生物化学自然发生的反应，而且还是热力学上属于熵增反应；再则，脂肪酸碳数不同的化合物以及脱羧反应形成不同碳数的烃类化合物所造成的热力学参数的差异都是比较小的数值，不值得苛求高碳数、高沸点、高难度实验测定每一个反应和每一个化合物的热力学参数。地质体中的时间漫长，反应速率参数可以不考虑。至于大庆油田油田水中检测出高含量（约 2000 mg/L）脂肪酸，笔者认为这不是化学反应式（1）不彻底的问题，而是碳酸盐已消耗殆尽，即缺少碳酸盐，属于质量作用定律范围的问题。

## 4　结论

石油成因体系的化学基础是一个综合的整体理论体系，这个体系用赋予一种数学概念的图——石油成因化学反应方程集示意图表达;属于该集合的元素由 4 个简明的化学反应式构成,各化学反应式具有特定的含义且相互密切关联。这样,可以一目了然其必然的、深刻的内涵。

根据化学反应平衡和质量作用定律的基本原理,确定碳酸盐对石油生成化学反应具有极其重要的作用。

(1) 碳酸盐与生烃化学反应中反应产物 $CO_2$ 发生化学反应生成可溶于水的重碳酸盐,从而有效、有力地保证了生烃反应的长期而稳定地进行。

(2) 碳酸盐的来源有两个:一是碎屑沉积碳酸盐;二是重碳酸盐的分解反应产生的碳酸盐沉淀,可称为自生碳酸盐。

(3) 重碳酸盐的来源是沉积古物源区的碳酸盐的风化溶蚀作用进入沉积盆地水体。

重碳酸盐分解反应生成的产物还有另外三个既特殊又极为重要的作用。一是二氧化碳作为光合作用的碳源,产生了能生成石油烃类的能转化为脂肪酸的糖类以及组成约占石油一半成分的石油非烃类、胶质和沥青质等。二是从水体中析出的碱土金属元素钙、镁等具有纳米级特征,为水体生物提供了丰富的物质来源。叶绿素-镁的作用等都是毫无疑问的巨大。三是二氧化碳在水生生物中的生物化学作用和生物代谢作用等,这种作用前人已有大量的肯定性结论,笔者则把这些成果提升到石油生成理论领域。

热力学理论和化学反应动力学理论以及质量作用定律的基本原理检验和证实了笔者所创立的石油形成化学基础的理论体系是正确的;一个湿封闭体系中的生物化学反应产生石油烃类,必然面临着许多复杂的物理化学难题,然而,这些基本的热力学、反应动力学、质量作用定律等理论解决了这些难题,揭示了石油形成的化学-物理化学机制。

根据石油成因体系化学理论分析,很自然地揭示出油田形成的奥秘是:碳酸盐对石油烃类生成反应以及油田形成都具有极其重要的作用,没有碳酸盐就没有石油;在沉积盆地,大量碳酸盐只能来源于盆地周边的陆地,给沉积盆地提供沉积物的被地质学家称为古物源的碳酸盐碎屑以及自生碳酸盐参与构建的湿封闭体系,是油田形成的决定性要素。

笔者所创立的这种理论是能够经得起大量石油地质实体检验的。世界上最大的加瓦尔油田就是最好的验证。笔者确定,加瓦尔油田的主力油源岩是侏罗系的"碎屑石灰岩",而非阿拉伯-美国石油公司高级地质顾问 Abdulkader Afifiji 认为的油源岩是侏罗系的"富含有机质的灰泥岩"。这一重大认识分歧的最终裁决将既是世界范围内对笔者的理论的最有力的检验,而且也将对世界地质科学产生极大影响。

笔者认为的其他检验事实有：大庆油田最好的油砂普遍是孔隙中无碳酸盐残留，同时又能在油田水中普遍检测到高含量的脂肪酸；长庆油田数以万计的薄片鉴定资料，优质油砂无碳酸盐胶结物。还有部分事例已在笔者发表的文章[17－19]中可以找到，而笔者所创立的有机生油理论铁证生油剖面毫无疑问地是这种理论的铁证[15]。

## 参考文献

[1] 何志高. 湿封闭体系综合效应成油论[J]. 石油实验地质，1988，10(4).

[2] 何志高. 论石油烃类气体成因[J]. 地质学报，1986，60(4).

[3] He Zhigao. A model of oil-generation in a water logged and closed system[J]. J Southeast Asian Earth Sci，1991，5(1－4).

[4] He Zhigao，Shen Liduo. Mudrock not being oil-gas source rock[C]. 30th International Geology Congress，1996.

[5] 何志高. 煤不成油的理论基础[J]. 石油知识，2002，(92).

[6] 何志高. 论干酪根热降解成油理论弊端[J]. 海相油气地质，1999，4(2).

[7] 何志高. 陆相石油与生油理论[J]. 石油知识，2003，(3).

[8] 何志高. 应用酸解气态烃研究碳酸盐生油岩[J]. 石油实验地质，1983，(2).

[9] 何志高. 关于石油成因理论发展历史若干问题的思考与评论[J]. 海相油气地质，2000，5(3－4).

[10] 何志高，樊文玲. 试用油气成因新理论讨论台湾油气勘探选区[J]. 台湾石油通讯，1988，(2).

[11] 何志高. 台西盆地油气组成与性质差异的成因分析[J]. 台湾石油通讯，1992，(2).

[12] 何志高. 天然气组成特点及成因机理分析[J]. 石油实验地质，1985，(4).

[13] 何志高. 煤成气量的热化学理论计算方法[M]//天然气勘探，北京：石油工业出版社，1986：200－203.

[14] 何志高. 从热力学理论讨论烃类演化[J]. 石油实验地质，1982，(1).

[15] 何志高. 中国“加瓦尔胚胎”与有机成油理论铁证生油剖面 一种新命名的岩石——“油泥灰岩”命名申述[J]. 石油知识，2011，(148).

[16] 张正斌. 海洋化学[M]. 青岛：海洋大学出版社，2004.

[17] 何志高. 关于沉积地球化学指标铁的还原系数[J]. 石油实验地质，1965，(1).

[18] 何志高. 红外光谱在石油地质应用中的几个问题[J]. 石油实验地质，1978，(1).

[19]) He Zhigao，Shen Liduo. Generation of petroleum hydrocarbon gases directly by microbioorganic fermentation[C]. The 30th International Geology Congress，1996.

# 6. 关于石油成因理论发展历史若干问题的思考与评论

## Thought and comment on some confused problems in the oil origin theory

何志高

（中国石化集团新星石油公司实验地质研究院）

**摘要**：对石油成因理论争论中的几个问题提出了以下见解：① 煤不成油，该结论可以由两种简单而有力的方法予以证明。② 无机成因学说的依据至今仍然相当欠缺，金刚石包体中的气体不是“石油类”物质而更像是“煤类”物质；原油、沥青的 Pb、Sr 同位素测试中因干扰因素很多而会影响测试精度，应予以充分注意。③ 有机成因理论中最为有力的证据是，在基本代表全球成矿时间和空间的 8 个地质年代统计中，世界上石油和煤、油页岩已知储量分布的曲线变化趋势非常相似。④ 泥（质）岩不是油源岩，特定条件下的砂质沉积物可以高度富集成油有机物。指出了在建立科学的理论时应注意的几个问题，尤其应注意石油类物质区别于其他物质的本质特征与特性。呼吁支持与应用湿封闭体系综合效应成油论，这一理论解答了有机成因理论中最难的初次运移问题，指出了碳酸盐岩窄而厚的相变过渡带为油气富集带。

**主题词**：石油成因；无机成因；有机成因；生油理论；发展史

在 20 世纪 60 年代，苏联著名学者 М. Ф. Давари 称石油成因是自然科学中最复杂的问题[1]。近半个世纪过去了，进入了 21 世纪，石油成因依然还是假说，还在激烈争论中，这种事实不能不承认 Давари 的上述评论有些道理。的确，石油这一种特殊的矿产资源，基于它的流动性、易变性、组成的复杂性以及赋存的特殊性，其成因研究经历了有机—无机—有机的争论；在有机成因占主导地位时，无机论者不乏其人，时起时伏，不少学者还改变自己的主张；在有机成因范畴内又有早期与晚期、干与湿之争；而多元论者则提出诸如早也生、晚也生，干也生、湿也生，无机也生、有机也生，甚至还有“半无机”成因、“部分无机、部分有机”成因之说，等等[2-7]。如果从俄国化学家罗蒙诺索夫1763年提出石油是煤干馏产生的观点作为最早的有机成因理论开始算起，至今已有200多年；如果从门捷列夫1876年提出无机成因假说从而开始石油成因理论的大争论算起，至今也已100多年。

毫无疑问，石油成因理论的争论，在科学上已经起到过、今后也必将进一步起到有助于逼近成因真理的重要作用，同时会对提高油气勘探效率产生巨大的影响[8]。

历史的经验值得注意，近一二百年的历史经验更值得注意。当前，我国也出现了石油成因新理论，一种崭新的、完整的理论体系——湿封闭体系综合效应成油论[9,10]；同时，一股新的无机成油理论、非生物成油论的热潮也正在兴起[4]。本文是在这样的科技创新的时代精神鼓动下，思考一些历史发展变化的问题。

## 1 煤不成油

1763年，俄国学者罗蒙诺索夫在他去世的前两年提出了石油是煤在地热作用下干馏所产生的，他是最早的有机成因论者，当然也是最早的煤成油论者。

罗蒙诺索夫从德国留学归国后一直在科学院工作，他的著作很多，虽然有“诗人”、“哲学家”之称，但他主要从事化学研究。他有著作《论地层》[11]，说明他对地质是有研究的，并且也有过煤焦油与煤干馏的实验研究基础。从当时的分析测试水平来看，煤在实验干馏和自然界中加热焦化所产生的煤焦油，其外表颜色、密度、状态很像石油，说明他的观点是有依据的。

然而，200年后的今天，我国还有一大批煤成油的研究者和拥护者，他们似乎还是没有走出罗蒙诺索夫的认识范围。更有甚者，把煤成油作为中国石油地质研究中的一种“创新”，一种理论上的突破与成就。

煤成油最根本的错误是没有分清楚煤焦油与石油的基本区别，混淆了煤与煤系地层、煤与含煤盆地、煤的沉积环境与盆地的其他非煤沉积环境这些有着本质区别的地质体、地质概念[10]。可以肯定地说，煤是不能成油的，只不过煤与油的关系是密切的，主要表现在地史分布、空间组合关系以及沉积相及沉积环境变化上，而不是煤在地热作用下直接生成并排出石油。

要证明这种结论是不难的。正是这样容易证明，才使得罗蒙诺索夫的煤成油理论在世界上200多年内没有大市场。笔者认为，证明的方法有两种。

第一种证明方法是:取任何一种煤矿(包括任何牌号煤、任何类型煤以及任何煤盆地或地区)中的瓦斯气,做一种最简单的气相色谱分析,检测烃类气体的含量;再取任何一种油田的伴生气(或油溶气),也做与上述相同的分析检测。这两类样品分析测试的结果显示它们之间有着极大的本质区别。瓦斯气中,乙烷及乙烷以上重烷烃($C_2^+$)含量很少很少,而石油气中的 $C_2^+$ 含量则很高很高。这是一种最快速、最经济的方法,完全用不着有复杂结构与构型的高分子甾萜烷的分析测试。浅显的道理是:瓦斯气就是由煤直接生成的气,没有石油气的特征;煤成气与石油气是两种截然不同的气体;瓦斯气是就地的原生气,它的组成保持煤化过程的全部产物的组成,而不会有显著的吸附、分馏、稀释、氧化-还原、水的溶解及其微生物的降解等可能改变其组成的变化因素;相对含量很高的 $C_2^+$ 烃类是石油及石油气的共同的本质特征,怎么可能煤生成油呢!

第二种证明方法是:到任何一个煤矿看看,观察煤体裂隙(包括微裂隙)中有没有石油?煤层上、下接触处及煤层横向上的非煤岩石接触处有没有石油?当然,最好是到吐哈盆地——“煤成油”的盆地去看,而且要看正好处于“生油门限”深度的煤。看过以后的结论是肯定的,没有石油。这是再浅显不过的证明方法。笔者看过很多煤矿、煤层,也在吐哈盆地看过岩心,结论是有依据的。只有解决了“初次运移”才可能讲“二次运移”,这是常识。

## 2 关于无机成因学说

据报道[2],法国化学家柏斯勒(Berthelet)是最早的无机成因论者。他于1866年提出了碳化物乙炔成油假说。然而,10年后的1876年,门捷列夫提出无机成油论,鉴于他创立了科学的元素周期表以及有很多高研究水平的成就,享有崇高的威望,因而后人总是将无机成油理论创立者的桂冠给予他。不过他也确实博学多才,他是一个对地学有研究的化学家,他当年所提出的煤的地下气化的主张,至今都有意义。另外,他也不赞成煤成油。正因为他在地学上的研究成就也很大,从而对无机成油理论的“后继有人”产生了深刻的影响。

笔者认为,就目前的资料来看,无机成油在理论上和实际中的依据还是相当欠缺的。在本文中仅仅讨论以下两个问题:

### 2.1 关于金刚石包裹体

金刚石包裹体测定表明,包裹体中普遍含有 $CH_4$, $H_2$, CO, $CH_3OH$, $C_2H_5OH$ 等,并被认为这是“石油类”化合物,是金刚石结晶时形成的化合物,从而得出石油类化合物可以在高温高压条件下由非生物的无机物形成[5]，并以此作为无机生油的直接证据。

首先，笔者认为，这类化合物不能认作是“石油类”化合物，它们与石油类化合物差异极大，无论从定性组成来看还是从定量组成关系来看均是如此，从元素组成来看亦如此，而且还可以认为这些化合物不是成油反应的反应物。这些认识，在当今已发展到对石油研究达分子级水平，以及相当精确的定性、定量分析测试技术的现代化水平，是显而易见的结论。

这类化合物为什么不能看成是“煤类”物质、“油页岩类”物质或者“生物类”物质呢？

根据 J. N. Bhaumik 资料(表 1)，不同牌号煤中的 O、—OH 基团占很大比例(引自文献〔12〕。因此将金刚石包裹体中的化合物称之为“煤类有机物”不是更合适吗?!

**表 1　煤的含氧基团组成**(据 J. N. Bhaumik 资料，引自文献〔12〕)

| 煤号 | 元素组成(%) | | | | | 含氧基团组成(%) | | | | |
|---|---|---|---|---|---|---|---|---|---|---|
| | C | H | N | S | O | —OH | :CO | —COOH | 其他 | 总计 |
| A | 79.4 | 5.6 | 1.9 | 0.6 | 12.5 | 6.0 | 1.6 | 0.8 | 4.1 | 12.5 |
| B | 83.0 | 5.8 | 2.3 | 0.5 | 8.4 | 4.2 | 0.5 | 0.2 | 3.5 | 8.4 |
| C | 88.0 | 5.2 | 2.3 | 0.5 | 4.0 | 1.2 | 0.2 | 0.0 | 2.6 | 4.0 |

### 2.2　关于 Pb、Sr、Nd 同位素

关于 Pb、Sr、Nd 同位素证据的研究近来愈来愈活跃，笔者认为有两个问题应予以充分注意并做出准确的评估，方可得出结论。一是原油、沥青中 Pb、Sr、Nd 的来源多种多样，非石油类矿物的干扰令人忧虑；二是这些金属元素的含量很低，分析测试本身的准确度、检出限以及试剂、器皿污染、灰化过程中的同位素分馏等，都会影响测试结果。

关于来源，至少有以下几种：

(1) 烃源岩中有机物与无机矿物；

(2) 输导层的矿物与水；

(3) 储集岩矿物与水；

(4) 开采过程混入钻井液、矿物、水与机械杂质。

表 2 是主要伴生元素为 Pb、Sr 的矿物，而这些矿物在烃源岩、储集岩中是普遍存在的，如文石，这种产于表生过程并终将转换成同质异象的方解石，根据有关资料[13]，它的第一位伴生元素即为 Sr，而第二位则为 Pb。可见 Sr 和 Pb 混入原油与沥青是相当容易的。

**表 2 主要伴生元素为 Pb 和 Sr 的矿物**（据文献〔13〕）

| 矿　物 | 主要伴生元素 | 矿　物 | 主要伴生元素 |
|---|---|---|---|
| 文石(霰石) | Sr 和 Pb | 重晶石 | Sr |
| 方解石 | Pb | 硬骨膏 | Sr |
| 碳酸锶矿 | Pb | 碳酸钡 | Sr |
| 斜长石 | Sr | 碳酸铅 | Sr |

## 3 有机成油铁证如山

众所周知,有机成油理论的证据很多,范围很广,研究程度很深,资料非常丰富,有地质方面的,有化学、生物、微生物方面的,等等。本文对此不作赘述。笔者认为,在这些证据之中,排在首位的是石油储量的地史频率分布与煤储量和油页岩储量的一致[14],如图 1 所示。从图 1 可见,8 个成矿地质时代基本代表了全球的成矿时间,石油、煤和油页岩储量分布变化曲线的变化趋势非常相似,煤-油关系极为密切是极其明显的客观事实。

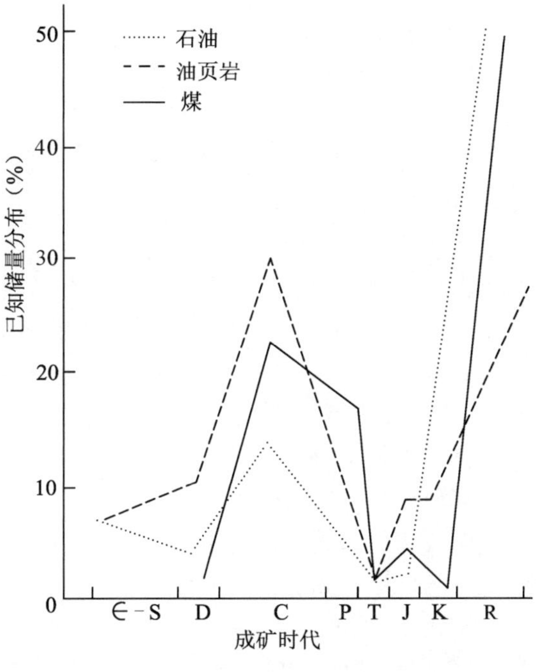


**图 1 世界石油、煤、油页岩已知储量按地质时代的分布**（据文献〔14〕）

煤是固定不变的固体矿产，在组成上以碳、氢元素为主，完全与石油一致，石油和煤都同属有机可燃矿产。此外，还有油页岩，就其中的有机物质成分而言，油页岩介于石油和煤之间，具有过渡性，在阐明有机成油方面具有旁证和提供多重证据的意义。

## 4 建立石油成因理论的几个要点

笔者认为，要建立科学的石油成因理论，以下几点必须认识清楚：

(1) 石油类物质具有区别于其他任何物质的特殊的、特定的性质和组成。换言之，不具备这些特性与组成特点的任何物质均不能被称为石油或石油类物质。

关于这些方面的资料十分丰富，无须赘述，现仅举例言之。

例如，在最简单的石油烃类气体组成上，一是湿度高，二是 $nC_4/iC_4$ 比值高，三是没有烯烃，而且这些特性均有限定的参数；在芳香烃上，大家都知道的布雷芳香烃结构分布指数——ASI 值(一种芳环上两类初级氢原子结构差异的吸光度比值)，对原油和"油源岩"而言，其值介于 0.8 ~ 1.4。这是典型的石油类物质的特征。

再例如，卟啉类化合物，它们在地质体中种类非常多，而在石油中却只有有限的几种，这些是特定的。

在历史上，除了关于煤能否生油的争论外，还有关于油页岩是否为生油岩、碳酸盐岩是否为生油岩以及有关"生油岩"指标的争论等。笔者认为，依据上述特殊、特定的原则，就不难得出种种石油成因理论研究中谁是谁非的结论。当然，这种原则只适应于未遭受后生变化的原油。如果一种遭受中等程度生物降解的原油的 $C_4$ 指标会改变，但是它的湿度与芳香烃结构分布指数是不会改变的。此外，对凝析油来说，它的重芳香烃含量会很低，芳香烃结构分布指数这种特征也就无从体现。总之，任何石油成因学说，一定要严格遵循石油物质的定性、定量组成的特征、特性。

(2) 泥(质)岩不是油源岩。

数十年来，在碎屑沉积区的油气勘探无一不是从泥(质)岩中寻找油源岩、评价油源岩；无论新区、老区，无论老专家还是实习的学生，都不例外。从教科书上，从指导思想上就认为油源岩必定是泥(质)岩。其实，目前业已发现，泥(质)岩不是油源岩。在泥(质)岩中寻找、评价油源岩是石油成因理论研究中的最大失误，当然也是地质上的一大误区，是地质学家的一大偏见。

回顾历史，"干酪根热降解成油"这种不科学的不符合地质客观事实的理论，之所以能够形成并盛行于世数十年，其原因除了上述错误外，就是先入为主的观念[15]。在60年代的巴黎盆地研究石油形成过程中，即干酪根成油理论创立之初，先从浅层泥（质）岩开始，分析其中的可溶有机物质和不溶有机质，再往深层采样、分析测试[6]。通过有机质数量与组成变化的从上到下的对比研究发现，随着深度的增加，有机质组成愈加接近石油，石油类物质逐渐增多，干酪根的元素组成随之变化，从而引出了从“不成熟”到“成熟”的概念以及“门限深度”、“门限温度”等一系列的“晚期成油”理论概念和理论依据。

（3）正确评价砂岩、砂质沉积物中的成油有机物质含量。

研究给出了这样的结论：并非所有的砂岩、砂质沉积物的有机物质含量都是很低的；相反，有些砂质沉积物、某些砂岩的“非现在时的”地史阶段，曾经含有极其丰富的成油有机物质，其量可以高达先期沉积的泥质沉积物的10倍。

根据资料，历史上曾有人提过砂岩生油的问题。但是，由于“现在时”研究的砂岩，其中有机物质不仅含量低，而且大颗粒的碳化碎屑占了很大部分，其生油能力很低。因此，将砂岩视为不可能作为形成油气藏的油源岩而加以否定。

湿封闭体系成油理论[9]首先发现了如图2所示的泥质沉积物中有机物质的重新分配作用，继而深入研究沼气发酵领域的微生物作用，不仅发现了发酵液中脂肪酸的含量与石油烃类气之间的脱羧规则，而且用实验的方法实现了直接生化反应形成石油烃类气及其 $C_6^+$ 更高分子的典型石油烃类，其结果如表3所示，给石油成因新理论以有力的支持。

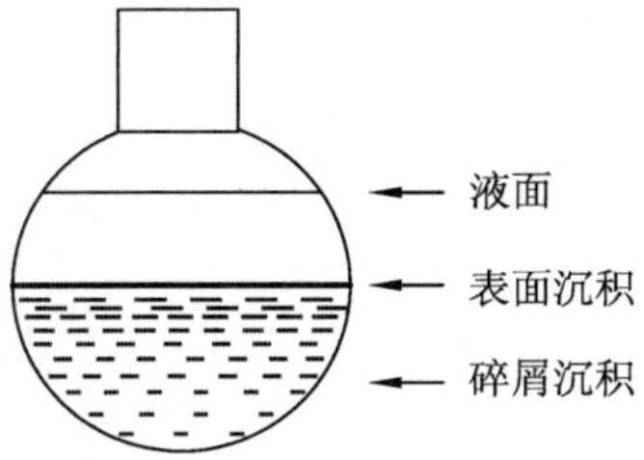


**图2　有机质富集在沉积物表面的模拟实验**

湿封闭体系成油理论没有初次运移、排烃难题，能解释任何石油地质客观事实。笔者相信，这一新理论最终必将得到世界石油界更广泛的认同与支持，并为全球的油气勘探开发做出特殊的贡献。

**表 3 微生物直接生化反应形成的石油烃类组成**

| 反应时间(h) | 烃类组成(体积分数,%) | | | | | | | |
|---|---|---|---|---|---|---|---|---|
| | $C_1$ | $C_2$ | $C_3$ | $iC_4$ | $nC_4$ | $iC_5$ | $nC_5$ | $C_6^+$ |
| 6 | 98.1 | 1.4 | 0.07 | 0.17 | 0.08 | 0.14 | 0.05 | * |
| 12 | 92.9 | 3.7 | 0.40 | 0.22 | 0.99 | 0.52 | 0.87 | 0.40 |
| 29 | 92.5 | 3.2 | 0.50 | 0.42 | 1.63 | 0.74 | 1.05 | * |
| 36 | 96.5 | 0.4 | * | * | 1.40 | 0.68 | 1.01 | * |
| 95 | 92.5 | 0.1 | 0.38 | 1.79 | 1.38 | 2.07 | 0.37 | 1.31 |
| 平均 | 94.5 | 1.8 | 0.27 | 0.52 | 1.10 | 0.83 | 0.67 | 0.34 |

* 值低于检测限(色谱定量分析)

(4) 碳酸盐岩地区石油地质具有特殊性及油气富集带。

笔者 15 年前发表了《窄而厚的相变过渡带沉积物富含石油——碳酸盐岩地区找油的重要领域》①一文后,一些研究者陆续肯定了碳酸盐岩地区的特殊性[16],但是在油源上仍然争论很多。笔者认为,必须用“湿封闭体系”这个理论来研究碳酸盐岩才能找到碳酸盐岩地区的油气富集带,才能避免盲目性,提高油气普查勘探效率,才能在成油理论的普遍性、统一性的指导下,阐述碳酸盐岩沉积物和碎屑沉积物各个不同的具有特殊性的地质体中石油的生成问题。

## 参考文献

[1] 佚名译. 关于石油地质和地球化学的理论研究方向[J]. 石油地质实验文摘,1963(1):3.

[2] 柯清水. 石油与石油工业[M]. 台湾台北:正文书局,1980.

[3] 陈文贤,朱炳泉,张景廉,等. 辽河断陷原油生成环境与演化[M]. 北京:石油工业出版社,1999.

[4] 陈淦. 油气无机成因说与油气成因二元论[J]. 石油知识,1998(4):11-12.

[5] 张子枢. 金刚石与油气结缘[J]. 石油知识,1995(1):8-9.

[6] Week LG. Origin, migration and occurrence of petroleum[M]//Petroleum Exploration, Handbook, New York: McGraw-Hill, 1961.

[7] Bailey Kascoe J, Alder FJ. Permo-carboniferous hydrocarbon accumulations, Mid-Continetn, USA[J]. *AAPG* Bulletin, 1983, 67(6).

① 何志高. 窄而厚的相变过渡带沉积物富含石油——碳酸盐岩地区找油的重要领域. 见:石油地质中心实验室测试汇编. 有机地球化学论文集,1986.

[8] 何志高. 发展石油成因新理论刻不容缓[J]. 石油知识,1995(4):36 - 37.

[9] He Zhigao. A model of oil-generation in a waterlogged and closed system[J]. J Southeast Asian Earth Sci,1991,5(1 - 4):456 - 469.

[10] 何志高. 再提"湿封闭体系综合效应成油论"[J]. 石油知识,1998(4):1.

[11] Brook JD,Maher TP. Acidic oxygen containing group sin coal[M]. Fuel, 1957,36.

[12] 四川省地质矿产局,华北地质研究所. 矿物原料分析方法[M]. 北京:地质出版社,1972:429 - 441.

[13] 西北大学地质系石油地质教研室. 石油地质学[M]. 北京:地质出版社,1979:33.

[14] 何志高. 论干酪根热降解成油理论弊端[J]. 海相油气地质,1999,4(2):50 - 56.

[15] 关士聪,演怀玉,丘东洲,等. 中国海陆变迁、海域沉积相及油气(晚元古代—三叠纪)[M]. 北京:科学出版社,1984.

本文原载于《海相油气地质》2000 年第 5 卷第 3 - 4 期

# 7. 论干酪根热降解成油理论弊端

## On the drawbacks of the theory of oil generation by thermal degradation of kerogen

何志高

(中国新星石油公司石油地质中心实验室)

**摘要**：石油成因是自然科学中一个复杂的难题。干酪根热降解成油理论起始于20世纪60年代，并逐渐被广泛接受。然而该理论存在如下弊端：(1) 不能解释“不成熟”原油的较大规模聚集；(2) 面临着排烃与初次运移难题；(3) 干酪根热解产物在组成上与天然原油(气)差异很大，二者可比性很差；(4) 油-岩对比可靠性差。总之，干酪根理论在很多方面不符合石油地质与地球化学实际。

**主题词**：油气成因；生油理论；烃类演化；干酪根演化；油气运移；烃类组分；热解产物；油源对比

干酪根热降解成油理论起始于20世纪60年代，70年代中期已发展为鼎盛阶段，此后一直指导着全球性油气勘探与油气地质的研究。然而，该理论的弊端也日渐突出地显露出来。

石油在地质体中的流动性、易变性、组成的复杂性以及赋存规律等特殊性，使得石油成因研究的一个多世纪，经历了无机与有机、早期与晚期、干与湿等理论上的长期争鸣，至今仍是假说。石油成因是自然科学中一个复杂的难题，是自然界的一种奥秘。石油成因无疑也是石油地质学的基础理论，这个问题不解决，许多问题不明，石油勘探事倍功半，解决成因问题犹如油气地质与油气勘探中的高屋建瓴，对干酪根成因理论弊端的研究，将会推动油气起源的研究。

## 1 关于"不成熟"石油

"不成熟"石油广泛地出现在各种文献中，尽管这种石油很早以前为人们所知。按照亨特(1982)的意见，世界上9%的液态烃储量、7%的天然气储量和40%的沥青是在成岩作用(早后生作用)时形成的。这种石油矿藏在各种类型的盆地中均有发现，它们的同生性不仅被石油和有机质的地球化学指标所证实，也被具体的地质环境所证实[1]。

L. R. Snowdon 与 T. G. Powell 曾描述了加拿大的波弗特-马更些第三纪盆地内不成熟石油是由陆相有机质形成的，其演化程度低，镜质体反射率($R_0$)为0.4% ~0.6%；O. K. Баженова[1]研究了库页岛东部的石油指出，无论石油或岩石中分散的有机质，均含有大量的生物标记化合物——甾烷、萜烷，根据乙基胆甾烷的成熟度指标离极限值很远，从而说明无论石油或沥青类都是不成熟的；在加利福尼亚州新生代盆地中，不成熟石油在600 ~2800 m深的地层内均有发现。O. K. Баженова[1]指出，深度2000 m以内的这种不成熟石油，其特点是胶质含量高、密度大、富含硫，其数量可观，可形成工业性油藏。

在中国，江汉盆地、百色盆地、济阳坳陷等地均发现了不成熟石油。江汉盆地广华寺构造的广33井原油，产层深度1917 m，属自生自储，$R_0$等于0.43%，为典型未成熟原油。该盆地板1井原油是在含膏盐地层中发现的具有工业价值的原油，产层位于610.23 ~658.5 m，密度为1.0086，硫含量为11.54%，可对比的相应油源岩的$R_0$为0.36%，富含生物标记化合物，在甾烷分析上仍保持着生物原始构型的优势，如$C_{27}$、$C_{28}$、$C_{29}$20R化合物，其中$C_{27}$甾烷$\beta\beta/(\alpha\alpha+\beta\beta)<0.1$、$C_{29}$甾烷$20S/(20S+20R)<0.3$、$C_{29}$甾烷$\beta\beta/(\alpha\alpha+\beta\beta)\leqslant 0.25$，均未超过A. S. Mackenzie等的成熟门限标准[2]，均是十分低的成熟度(据傅家谟等资料)。

众所周知，干酪根热降解成油论是指细粒沉积物中的有机质随着埋深作用经受愈来愈高的温度，从而使干酪根发生并逐渐增强降解成烃作用，其形成大量烃类的时期就是油母质生油成熟期。依此机制，不言而喻，对一个具体含油盆地来说，只有进入成熟期，才有大量液态油可言，而不成熟原油占有很大比例的油田，如江汉油田、辽河油田就不应该存在。至于干酪根理论将成熟前的少量烃类作为对不成熟油来源的解释，除了存在后文所讨论的排烃难因素外，就一个油田的原油量来说，成熟前的少量烃类无论如何形成不了一个年产数十万吨乃至数百万吨油田的三分之一以上的原油量，更无法形成占世界储量9%的液态烃、7%的天然气、40%的沥青。因此，"不成熟"原油本身就是对干酪根成油理论的一种批判、一种否定。

有一种观点，例如 Snowdon 和 Powell 认为不成熟原油的物源是树脂类物质，这类物质不仅数量多，而且可以低温生油。然而，重要的问题不是能否生油，而是比生油更难解释的初次运移难题。前已述及，"不成熟"石油胶质含量高、密度大、含轻烃和气态烃组分少，毫无疑问，这种油的黏度大，且是非牛顿液体，若无很大的压力差和很高的温度，这种石油是不可能从"生油岩"中运移出来的。

## 2　关于排烃与初次运移

干酪根热降解成油理论从诞生那时起，就面临着一个排烃与初次运移难题。直至发展 20 多年之后的今天，它仍然是一个致命性的障碍，Tissot 前不久还承认，石油中的重组分的排出是一个难以解答的问题。

其实，初次运移尚未得到最起码的直接证明，纯属推断。

推断的一种解释是单分子的排出与运移，即石油以气体分子状态从"生油岩"排出并(初次)运移至储集岩。这种观点不可思议。人们在讨论气态烃的扩散与运移时，还承认有分子大小、排出能力大小之别，何况其分子量、分子极性比甲烷、丁烷大千百倍的胶质与沥青质了。可以肯定地说，没有高真空、高温条件，很多重烃分子、几乎全部胶质与沥青质分子都不可能以分子气体状态出现。

从事过原油蒸馏实验的人都知道，蒸馏原油，包括恩氏蒸馏、精密蒸馏、减压真空蒸馏绝不可能蒸馏出原油中的沥青质。而当蒸馏温度达到约 600 K (1 个大气压下)时，原油高分子的分解反应就明显地发生了。岩石热解分析亦表明，人们还没有取得过将岩石中的沥青质通过加热(包括居里点方法)，使之从岩石中释放出来的实验结果。在温度为 600 K 并有流动吹赶的载气条件下，收集到的产物仅仅是蒸发烃。当温度再增高，其产物即胶质、沥青质和干酪根的热解烃了。以正构一元脂肪酸为例，当达到脂肪酸的沸点温度时，脂肪酸随即开始分解。胶质和沥青质分子在高温条件下，C—N、C—S、S—S、N—O、O—O 键以及芳族化合物侧链 β 位键等化合键必然断裂。何况，处于地层深处的"生油岩"根本不是低真空或负压，而是高压，这就更增加了重组分汽化成气态分子的不可能性。增大压力对液态分子的汽化过程来说是热力学中的一个巨大熵减效应。因此，初次运移的方式不可能是以气态分子形式进行的[3]。

有人主张排烃与初次运移是液态方式。例如，I. Özkaya 在《用计算机模拟科威特原油的初次运移》[4]一文中的结论是：原油初次运移很可能是一个“单相(separate phase)过程”。

如果初次运移真是以液态方式，那么，在系统的地层岩心剖面中将可以广泛地见到含油饱和度相当高的油源岩，或者沉积有机质从不成熟到成熟阶段的含油状态的过渡变化。即油源岩油饱和度的显著变化。实际上，除了个别零星数据外，尚未见到有实测的油源岩含油饱和度的系统而可靠的变化资料。笔者还观察过许多油田及其钻井岩心，从未发现过这种高饱和度的油源岩，也未发现过渡变化，甚至没有发现油味变浓的嗅觉反应。

参照 Özkaya[4]的方法，对中国油田进行对比研究，发现他提出的科威特油源岩可转换成油气的有机质数量为 1%（体积），这个数量与中国各大油田的实际情况相距甚大，该值是中国多数油田油源岩($200\times10^{-6}$)的 20 倍，是大庆油田最佳油源岩($1000\times10^{-6}$)的 4 倍，是柴达木盆地旱 1 井成熟生油岩井段($65\times10^{-6}$)的 60 倍。按照这些资料，并且设定有效孔隙度为 5%，岩石密度为 2.5，油密度为 0.8 等参数进行估算，则中国大多数油田的油源岩的含油饱和度为 1.3%，大庆油田为数不多的最佳油源岩为 6.4%，旱 1 井为 0.4%。很明显，这些油源岩的油饱和度均很低，都远远低于 Özkaya 所应用的油源岩含油饱和度应该达到发生初次运移的临界值。

有人主张排烃与初次运移方式是以油水混相（或油水相互溶解）的方式进行的。其实，这种观点也是一种主观臆测。

原油在高压釜中脱水的实验表明，在高压高温条件下，油水只有分离的性质，而无混相的性质，正是依靠这个原理而得以使原油脱水，使已经形成的为量不多的油水乳状体得以解体。况且，水的沸点相对于油类中主要的烃类来说是相当低的，即使黏土矿物脱出的水分子也会蒸发逸散。从此种意义而言，“生油岩”在处于生烃高峰时所形成的烃类难以与水形成乳状体。即使可以形成，这种乳状体也是不稳定的。因为，当乳状体一旦脱离“生油岩”发生初次运移，环境温度就改变。温度一改变，原油的一系列性质就会发生变化。例如，加利福尼亚原油，100 ℃时的动力黏度为 50 mPa·s，当温度降低至 50 ℃时，黏度变到 1200 mPa·s；而 100 ℃的纯水，动力黏度为 0.28 mPa·s，50 ℃时为 0.55 mPa·s（根据 P. Ungerer 等）[5]。就是由于这种温度变化所造成性质差异的变化，油水乳状体方才被解体。

此外，很有意义的是，我们在地质体中所看到的含油砂岩，其含油部分与非含油部分的界线泾渭分明。这种油水分离的现象也是对油水乳状体理论的否定。

## 3 热解产物与油气在组成上可比性差

干酪根(或碳质泥岩)热解产物在组成上与天然原油、天然气的组成差异很大,二者可比性差,这是干酪根热降解成油理论的一个重要弊端,其表现主要有以下几个方面:

### 3.1 气态烃的干(湿)度

根据大量的热解模拟实验资料,干酪根热解产气,甲烷在气体中的含量很低,即 $C_1/C_2^+$ 比值很低,乙烷及其他重烃含量很高,湿度很大。相反地,天然气(包括石油伴生气)中,甲烷在气体中的含量很高,即 $C_1/C_2^+$ 比值很高,乙烷及其他重组分含量低,湿度也较低。二者之间此种差异已是众所周知的事实,无须赘述。

有人认为,这种差别的主要原因是"天然气从油源岩到储层的运移过程中发生的轻重分异"[6]。这种观点貌似有理,但实际上是谬误的。

气态烃从"油气源岩"向储层运移,可能有轻重分异,但这种分异只能是运移的先后次序。甲烷首先运移,随后是乙烷、丙烷等,正如色谱分析保留时间上的差异那样,但是这种可能性是极个别的、局部的、瞬间的现象。因为,这种保留时间及各组分保留时间之差值相对地质体中形成工业油气藏所需时间来说是极其短暂的,而且油气藏的气从源岩经过输导层进入储层的过程并不像色谱分析那样"一次进样",也不会是气源岩先生成甲烷,先运移完,然后再生成乙烷。因此,对形成工业油气藏来说,轻重分异不会发生。反过来说,假如发生轻重分异,那么在地质体中一定会找到数量相当多的、几乎一半以上的以乙烷、丙烷为主组成的天然气藏。可是,实际上地质体中并无此种气藏。

此外,"轻重分异"观点难以对丁烷两种异构体之间的关系作出解释:正丁烷的沸点比异丁烷高 12 ℃,熵 $S^0$比异丁烷高 3.7 cal·(deg·mol)$^{-1}$,色谱分析中比异丁烷保留时间长。倘若初次运移确实有"色层效应",那么残留在岩石中的重组分必然多,天然气中异丁烷一定要高于正丁烷。可是实际上,天然气中正丁烷普遍高于异丁烷。显然没有初次运移的轻重分异。

造成干酪根热解产物与油气在组成上干湿度差异的原因是两种化学反应,反应物与反应类型不同,因而反应产物不同。因此,石油烃类气不是由干酪根热降解而形成的。

### 3.2 气态烃中正、异构丁烷含量差异

一种化学反应,其反应产物的同分异构体之间的含量关系是判别该反应

的很重要的指标，是研究该反应性质、条件的重要依据。石油烃类气中的丁烷异构体是饱和烃类化合物中最简单的最低碳数的异构体，无疑是研究生成石油的化学反应的最基本的对象。

然而，根据天然气的地球化学特征，在讨论这个同分异构体的对比时，要排除天然气形成过程和油气藏保存条件对气态烃中正异构丁烷的影响。根据各种资料，影响正异构丁烷含量的最大因素是生物降解。其他的影响因素诸如演化程度（热稳定性）、沸点高低和分子极性对运移过程产生的分异作用等因素均较小。很多研究者忽视了这个问题，从而导致认识上的混乱。因此，这里特别需要强调对比的前提是，要消除生物降解的影响。

A. T. James 和 B. J. Burns[7] 对澳大利亚和加拿大一些遭受生物降解的天然气进行了相当详细的系统研究。表 1 是根据他们的有关资料编纂而成的，其生物降解程度由原油气相色谱及各种气态烃的碳同位素等资料确定。

**表 1　生物降解正、异构烷烃相对含量（mol%）**

| | 样　品 | | |
|---|---|---|---|
| | Elmworth<br>加拿大阿尔伯塔<br>（2232 m） | Sirius-1<br>澳大利亚滨海岸<br>（3019 m） | Eendracht-1<br>澳大利亚滨海岸<br>（2651 m） |
| 异丁烷 $iC_4$ | 0.45 | 0.12 | 0.09 |
| 正丁烷 $nC_4$ | 0.42 | 0.06 | 0.01 |
| $iC_4/nC_4$ | 1.10 | 2.00 | 9.00 |
| 生物降解程度 | | 低——→高 | |

从表 1 可见，随着生物降解程度的增加，正丁烷相对异丁烷损失增加，$iC_4/nC_4$ 比值增大，生物降解对丁烷异构体比值的影响是很大的。

根据美国矿务局早年对 2100 多个气样的统计，几乎所有的气体正丁烷含量高于异丁烷。

笔者对江苏油田（下第三系）的油溶气组成和性质进行了系统、详细的研究，确定用高压地层原油取样中脱附出来的烃类气体，在消除了压力因素、油藏保存条件优劣以及可能的生物降解因素后，与原油本身在成因上关系最为密切的石油烃类气体作为统计的基础，总共统计约 40 口井的资料，其中三垛组（$E_3s$）$iC_4/nC_4$ 比值平均为 0.355，戴南组（$E_3d$）该比值平均为 0.309，最能代表石油烃类气体原始组成的真 80 井，其戴南组油层深度为 3061 m 的油溶气，其 $iC_4/nC_4$ 比值为 0.287。

热解模拟试验的资料表明，$iC_4/nC_4$ 比值介于 0.4 ~ 2.0，生烃的主要时期,该比值约等于1.0。这个数字无疑与0.287差距甚大,显示了热解模拟实验结果与天然地质体中石油烃类气之间的重大差异。

在$iC_4/nC_4$比值上,对干酪根热降解成油论及其所谓的“油源岩”、“气源岩”更有力批判的事实是,这些岩石中的吸附烃$iC_4/nC_4$比值与石油烃类气中$iC_4/nC_4$比值差别很大。

本文所应用的吸附烃资料是由一种可靠的分析技术提供的。该方法[8]可能有其他缺陷,但是$iC_4/nC_4$是相当准确的。它将岩石颗粒($d=5$ mm),装入一特别的金属圆筒,加水浸泡、密封,从顶部抽出部分水,形成负压,启动搅拌切削装置,使气体脱附出来并贮集于顶部空间,然后取出样品进行色谱分析。

表2是中国某含油气盆地不同构造中的4口井52个泥岩和煤样品的吸附烃$iC_4/nC_4$比值资料。

**表2 中国某含油气盆地不同构造中泥岩、煤样品中吸附烃$iC_4/nC_4$比值**

| 井号 | 井深(m) | 泥岩(包括煤)样 | 泥岩(包括岩屑)样 | 岩心(泥岩)样 | 煤样 |
|---|---|---|---|---|---|
| Z—1 | 1460~3600 | 1.29(13) | 1.20(12) | 1.14(7) | 2.35(1) |
| B—1 | 1780~4022 | 1.44(9) | 1.30(8) | 1.04(6) | 2.56(1) |
| B—2 | 2000~3780 | 1.51(13) | 1.48(12) | 1.22(7) | 1.88(1)* |
| D—1 | 1600~3870 | 1.88(17) | 1.78(14) | 1.62(3) | 2.71(3) |
| 平均值 | | 1.56 | 1.47 | 1.20 | 2.44 |

注:括号内数字为样品数;*为沥青煤。

从表2可见,4口井52个通常认为“油源岩”的吸附烃不具备石油烃类气的特点,$iC_4/nC_4$比值距离0.287甚远,这种差别不是演化程度差异所致。因为4口井的$iC_4/nC_4$比值都没有随井深度变化而有规律地变化,4口井中5个煤样的镜质组反射率($R_0$)与$iC_4/nC_4$比值无关系。这种差别也不是运移分异(包括色层效应与轻重分异)所致,更不是生物降解程度差异所致。因为运移出“油源岩”的烃类气理应更易于遭受生物降解,而生物降解是选择性首先消失正丁烷,从而导致$iC_4/nC_4$比值增大,结果只会更远离0.287。

毫无疑问,石油烃类气体不是干酪根热降解形成的,它的源岩不是细粒富含有机质的泥质岩。

### 3.3 烯烃问题

凡是热解模拟实验(除有水条件实验),均有一系列的大量烯烃产生,这是众所周知的事实;同样,凡是原油和天然气,烯烃含量很低或者不能检测出,这也是公认的事实。无疑,这是一对矛盾。在有水存在条件下的热解实验，其产物中烯烃含量很低。例如，Lewan(1979,1983)、Winter(1983)等指出，富含有机质页岩经加水热解可以生成类似于天然原油的产物，排出的油聚集在水面上，它不含可测定的烯烃类。然而，这种实验结果却不能从根本上解释上述矛盾的实质。因为沉积物埋深进入“成油门限”以后，岩石中的游离水早已殆尽。黏土矿物受热所释放出来的水量也难以形成有水的热解介质条件。

烯烃，例如乙烯的环化反应生成环己烷，乙烯加氢生成乙烷，根据反应的热力学计算，其生成自由焓变化值分别为 $-172.57\ \mathrm{kJ\cdot(reaction)^{-1}}$、$-101.01\mathrm{kJ\cdot(reaction)^{-1}}$，反应均可以自发地进行，并且地质体中的高压条件有利于削弱这种反应的熵减效应，使反应更易于进行。通过这种热力学性质分析，也许可以解释天然气中无烯烃成分的事实。

然而，最基本的事实是，在大量的各种地质体的各类样品(包括由不成熟到成熟过渡带的样品)中均未检测出显量烯烃。笔者研究了某盆地不同构造的4口井岩心脱附气，无一样品有烯烃。假如的确存在烯烃环化与加氢反应，那么这种由热解所产生的烯烃按理是可以被检测到的，因为乙烯的环化，例如形成环己烷的反应是三分子反应，尤其在高压条件和低渗透率的泥质岩中，其反应速度不会很快，应该在泥质岩的脱附气中能够检测到。在检测灵敏度很高的气相色谱中没有检测出，则完全可以判断不含烯烃。因此，在烯烃问题上可以得出如下结论：在地质体中不存在那种干酪根热降解理论所定义的“生油岩”与成油机制。

### 3.4 关于其他组成成分

热解模拟实验的产物，如干酪根的热解、沥青质(包括原油沥青质与岩石抽提物的沥青质)热解产物，其中所含的多系列化合物为天然原油中所没有的，或者在含量上相差很大。例如，苯、萘、蒽等苯环上无取代基的芳烃类化合物在热解产物中的含量远远高于原油。就苯对比而言，二者的浓度比大于100，Ⅲ型干酪根热解产物中苯的含量特别高，二者浓度比更高。那种把中国东海石油归属为Ⅲ型干酪根成因的观点，恐怕更难解释原油与热解产物的巨大差异。

酚、甲基酚、甲乙酚等类化合物在热解产物中的含量也要远远地高于原油中的含量。这类化合物在很多原油与很多石油类气体中均未被检测出。

此外，芳烃类化合物总量在陆相原油中含量很低，例如大庆原油中约为10%，然而陆相地层中的干酪根热解产物芳烃含量可达20%，Ⅲ型干酪根更高[9]。如此矛盾种种，不作赘述。

## 4 油-岩对比可靠性差

历来,油-岩对比及确定油源岩的研究是石油地质、地球化学及油气勘探开发中基本的,也是最重要的一个问题。该问题从普查勘探见到油的那天起就开始研究、开始争论,往往到油田开发进入末期还在争论不休,各持己见。塔里木盆地沙参 2 井是 1984 年 9 月出油的,其油源观点有五六种之多,至今众说纷纭。这种情况本身在一定意义上就说明了干酪根热降解成油理论的不可靠性、不科学性以及某种弊端。

很多油-岩对比研究十分牵强附会。例如,江苏油田的原油产层主要产量来自下第三系戴南组($E_3d$),由于该地层“生油条件”差,只好将“油源岩”定成下部的相距有近 300 m 的阜宁组($E_2f$),因为阜宁组是一套厚的暗色地层,富含有机质,“生油条件”好。于是就寻找各种断层,并假设阜宁组所生成的石油通过这些断层向上覆的戴南组运移。其实,大量的戴南组和阜宁组的原油对比研究表明,戴南组原油与阜宁组原油是两类组成和性质各具不同特点的原油,它们之间没有运移上的差异,戴南组原油绝不是来自阜宁组。

我们从柴达木盆地旱 2 井的全部地球化学分析资料[10]可以看到,从采样井井深 195 m 开始至井深 6018 m 的 34 块样品,有机碳含量最高 0.23%,处于“成熟”阶段的上第三系(井深 3350.8 ~ 5454 m)有机碳平均含量仅 0.17%,烃含量平均 $66\times10^{-6}$,抽提物中饱和烃含量平均 32.4%。这些基本的参数表明,旱 2 井全部已分析的“生油岩”均不够一般所公认的“油源岩”的资格,然而旱 2 井在井深 5475 ~ 5522 m 中却获得了原油。该油样在损失了很多轻烃部分之后,饱和烃含量达 64.75%,饱和烃中有一定数量的 $C_{22}$ 以上的石蜡烃,且明显高于相应岩石抽提物中高碳数石蜡烃的含量。由此可见,这种油-岩对比及油源岩的可靠性实在太差。

再举中国东海某一油气构造中的油-岩对比的实例,从生物标记化合物上说明油-岩的不可对比性。

图 1 是岩石抽提物和原油的饱和烃色谱-质谱分析中 217 m/e 离子色谱图的对比。图中的峰号含义见表 3。图 1(b)、图 1(c)、图 1(d)三张谱图选自 13 个“生油岩”,作为“生油岩”的典型代表。图 1(a)中的 Oil(3792 m)谱图是选自 7 个原油样中甾烷含量较高、离子强度较大的一个原油样。

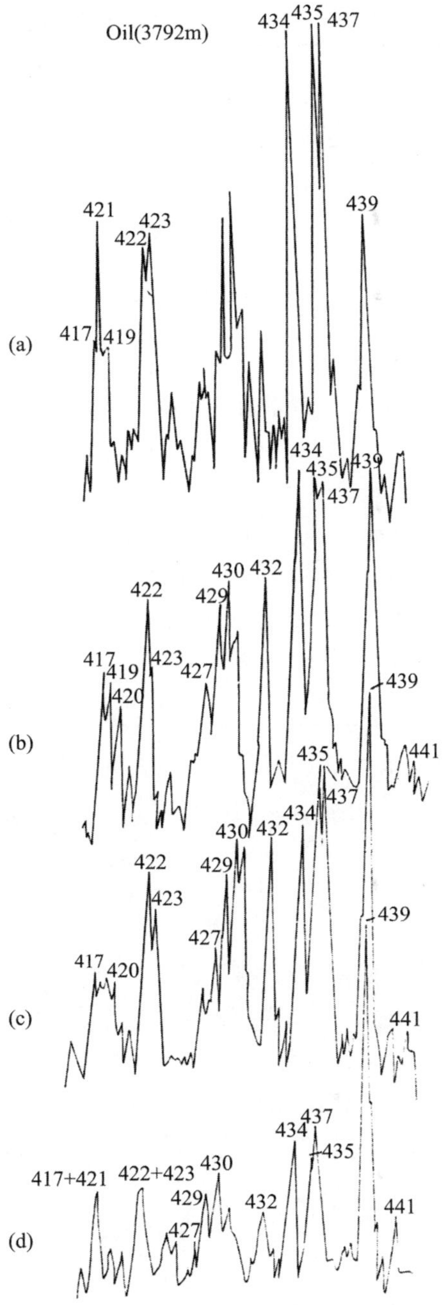


图 1 岩石抽提物和原油饱和烃色-质谱分析 217 m/e 离子色谱图

（图中峰号所示的化合物见表 3）

**表 3 图 1 中色-质谱分析谱图中峰号所示的化合物**

| 峰号 | 分子式 | 分子量 | 化合物名称 |
|---|---|---|---|
| 417 | $C_{27}H_{48}$ | 372 | 5α(H),14α(H),17α(H)-20S-胆甾烷 |
| 419 | $C_{27}H_{48}$ | 372 | 5α(H),14β(H)-20R-胆甾烷 |
| 420 | $C_{27}H_{48}$ | 372 | 5α(H),14β(H)-20S-胆甾烷 |
| 421 | $C_{29}H_{52}$ | 400 | 13β(H),17α(H)-20R-24-乙基重排胆甾烷 |
| 422 | $C_{27}H_{48}$ | 372 | 5α(H),14α(H),17α(H)-20R-胆甾烷 |
| 423 | $C_{29}H_{52}$ | 400 | 13α(H),17β(H)-20S-24-乙基重排胆甾烷 |
| 427 | $C_{28}H_{50}$ | 386 | 5α(H),14α(H),17α(H)-20S-24-甲基胆甾烷 |
| 429 | $C_{28}H_{50}$ | 386 | 5α(H),14β(H),17β(H)-20R-24-甲基胆甾烷 |
| 430 | $C_{28}H_{50}$ | 386 | 5α(H),14β(H),17β(H)-20S-24-甲基胆甾烷 |
| 432 | $C_{28}H_{50}$ | 386 | 5α(H),14α(H),17β(H)-20R-24-甲基胆甾烷 |
| 434 | $C_{29}H_{52}$ | 400 | 5α(H),14α(H),17α(H)-20S-24-乙基胆甾烷 |
| 435 | $C_{29}H_{52}$ | 400 | 5α(H),14β(H),17β(H)-20R-24-乙基胆甾烷 |
| 437 | $C_{29}H_{52}$ | 400 | 5α(H),14β(H),17β(H)-20S-24-乙基胆甾烷 |
| 439 | $C_{29}H_{52}$ | 400 | 5α(H),14α(H),17α(H)-20R-24-乙基胆甾烷 |
| 441 | $C_{30}H_{54}$ | 414 | 4-甲基-24-乙基胆甾烷 |

从图 1 可见,$C_{27}$至$C_{29}$共 12 种不同构型的规则(正常)甾烷,油-岩是不可比的,它们之间无共性、无任何相似之处可言。

## 5 结论

本文从“不成熟”石油、油气的初次运移、热解模拟实验产物、油-岩对比四个方面讨论了干酪根热降解成油理论的多种弊端,其中每一部分都有小结,都明确地指出了工业(商业)性石油不是干酪根热降解形成的,干酪根理论在很多方面不符合石油地质与地球化学实际。笔者认为争论了一个多世纪的石油成因,其主要通病是人们总是将细粒沉积物(泥岩)作为主要油源岩,从而导致人们在石油成因研究上走着曲折的道路。湿封闭体系综合效应成油论[11,12]可能会为最终解决石油成因理论开辟一条新的途径。

## 参考文献

[1] Баженова ОК. Образование Нефть на Неьольших Глубинах[J]. Геолоzuя Нефmь u Газа, 1990(7):2-7.

[2] MacKenzie AS, Maxwell JR. Organic maturation studies and fossil fuel exploration [J]. 1981:239-254.

[3] 何志高.热力学理论讨论烃类的演化[J].石油实验地质,1982,4(1):1-9.

[4] Özkaya I. Computer simulation of primary oil migration in Kuwait. Journal of Petroleum Geology 1991,14(1):37-48.

[5] Проскуряков ВА. Химия Нефть и Газа, Извдателвство. Хuмuя Ленинград, 1981.

[6] 李执,朱美茜,毛弘毅.源岩热解气态烃的碳同位素特征及其在气原对比中的应用[J].石油与天然气地质,1990,11(3):247-259.

[7] James AT, Burns BJ. Microbial alteration of subsurface natural gas accumulations[J]. AAPG Bull, 68(8):957-960.

[8] 詹智玉.岩石与天然气中 $C_1$—$C_9$ 烃类分析方法的研究[J].石油实验地质,1990,12(3):326-332.

[9] 何志高.松辽及其他陆相盆地生油岩地球化学的基本特征[J].石油实验地质,1981,3(3):197-202.

[10] 黄第藩.陆相有机质演化和成烃机理[M].北京:石油工业出版社,1984.

[11] He Zhigao. On the origin of petroleum hydrocarbon gases[J],Acta Geologica Sinica, 1986,60(4):57-65.

[12] He Zhigao. A model of oil-generation in a waterlogged and closed system[J]. J Southeast Asian Earth Sci, 1991, 5(1-4):456-469.

本文原载于《海相油气地质》1999 年第 4 卷第 2 期

# 8. On the drawbacks of the theory of oil generation by thermal degradation of kerogen

He Zhigao

(The Central Laboratory of Petroleum Geology, Ministry of Geology and Mineral Resources, Wuxi, Jiangsu, China, 214151)

**Abstract**: The paper discusses the drawbacks of the theory of oil generation by thermal degradation of kerogen. The principal arguments are as follows. In view of the theory, (1) No reasonable interpretation is available for the origin of "immature" oil; (2) Primary migration is a fatal problem; (3) A poor correlation between pyrolysed products and petroleum constitutions; (4) Oil/source correlation is not reliable.

The paper is a general review, and will be followed by a series of detailed studies on the subject.

**Key words**: oil generation by thermal degradation; theoretic drawback; kerogen

## 1 Introduction

Ever since its establishment in 1960s and prime in middle 1970s, the theory of oil generation by thermal degradation of kerogen has been guiding the study on the global petroleum exploration and geology. However, its drawbacks manifest themselves as time goes. The author attempts to analyse them in the paper, which is intended to be an introductory work of a systematic study.

Over a period of a century, the genetic study of petroleum generation gave rise to much controversy about inorganic and organic, early and late, dry and wet origins, but remains a hypothesis to the date, due to various factors, concerning fluidity, variability, compositional complexity and accumulative peculiarities.

Petroleum origin is a complicated tough problem in natural science, as well as a basic theory in petroleum geology. Many questions will not be clear, and more efforts will be taken for petroleum exploration and development with less results, before this problem is solved. Therefore, in order to have an overview in petroleum geology and exploration, it is necessary to take it into first consideration. The author believes that the study about the theoretic drawbacks will further the study of petroleum origin.

## 2 About "Immature" Oil

The concept of "immature" oil is well known in very early time. Of the total reserves in the world, 9% liquid hydrocarbon, 7% natural gas and 40% bitumen are formed during the process of diagenesis, according to Hunt (1982), "Immature" oil accumulations have been found in various basins. Their syngenetics is confirmed not only by geochemical indicators of petroleum and organics, but also by specific geological environments.

L. R. Snowdon and T. G. Powell used to describe the "immature" oil in Beaufort (1979), Canada. After a study on the oil of Sakhalin, O. K. Баженова suggested that the dispersed organics, whether in petroleum or in rocks, contain large amount of steranes and terpanes, which are far from maturity based on the maturity indexes of ethyl cholestane. In China, immature oil is found in Jianghan Basin and Jiyang depression etc. According to the data of Fu Jiamo, the crude oil in well No. 1 of Jianghan Basin is of commercial value, discovered in a salt-bearing layer. Its specific gravity is 1.0086, sulfur content is 11.54% and $R_0$ for a correlative corresponding source rock is 0.36%. The steranes in it still have a preponderance in primitive biogenic configurations, for instance, $\beta\beta/(\alpha\alpha+\beta\beta)<0.1$ for sterane $C_{27}$, and $\beta\beta/(\alpha\alpha+\beta\beta)\leqslant 0.25$ for $C_{29}$, and $20S/(20S+20R)<0.3$ for $C_{29}$, all of which do not exceed A. S. Mackenzie's standards of maturation threshold.

According to the theory of kerogen, large quantity of liquid oil can exist only when maturation is reached for a specific petroliferous basin. But this should not occur for those oil fields in which immature oil holds large proportion, such as in Jianghan and Liaohe fields in China. As for the small amount of hydrocarbons generated before the threshold is interpreted as the source of immature oil and the concept of an early generation from resin, it seems that the most important problem is the difficulty of primary migration. Immature oils are characteristic of high molecular weight and polarity, and they should expel from fine-grained sediments with very strong absorption, incredible to those who have some laboratory experience. Secondly, there is a question of quantity, for the small amount of hydrocarbons generated before the threshold can not formed one third of the crude oil in a field of 100000 ton or 1000000 ton annual outcome whatsoever; and even more is the case for the formation of 9% liquid hydrocarbon, 7% natural gas and 40% bitumen of world's total reserves. Therefore, "immature" oil itself is a criticism on the theory.

## 3 About the Question of Primary Migration

Ever since its establishment, the theory has been faced with the difficult problem of primary migration, and remains a fatal barrier after 20 years' development. Actually, to a great extent, primary migration is an inference only, and has not been proved by distinct evidences.

One hypothesis about it is that oil is in a state of gaseous monomolecular, which migrates into reservoirs after its expulsion from source rocks. This is reasonable in the case of gas and light oil, however, it does not account for moleculars of heavy hydrocarbon, nearly all are colloid and asphaltene, because it is contrary to physical and chemical facts. In an experimental distillation of crude oil, asphaltene does not gasify even if it is proceeded under a reduced pressure ( $-100$ mmHg) and a high temperature ( $>700$ K). In fact, the decomposition of high moleculars starts before the temperature is reached. A pyrolysis of asphaltene (including Curie point method) shows that asphaltene is unstable and the reaction is not reversible. Normal fatty monoacid will decompose as soon as it reaches boiling point. Therefore, the colloid and asphaltene in crude oil will not expel in a state of gas molecular from a source rock under any geological conditions. As for the vaporization of heavy hydrocarbons, the formational high pressure resulting in a very high entropy reduction adds more difficulty to the process.

Some advocate that primary migration occurs in a liquid mode, for example, Ozkaya in his recent paper (1991), draws a conclusion that the primary migration of crude oil is most probably a process of a "separate phase".

If that is the case, then, source rocks with high oil saturation should be found in an continuous formation core section; or there should be an extensive observation of the significant change in oil saturation from immature to mature stages. Unfortunately, except for few fragmentary data, the systematic and reliable data about oil saturating variation from any observed source rocks are not available so far. The author investigated many drilling cores from oil fields, and has never come across a source rock of high oil saturation (not including fractured mudstone reservoirs), and none of the case enconcentrated with increasing oil odour.

Refer to Ozkaya's method, a correlation with oil fields in China is made. The results show that the amount of petroleum transformable organics (1% volume) suggested by him is much larger than the actual cases of the major oil fields in China. His values are 20 times as much as those of source rocks (200 ppm) in most of oil fields in China, respectively, 4 times of the best source rocks (1000 ppm) in Daqing oil field; 60 times as much as mature source rocks (65 ppm) in Han No. 1 well in Chaidamu Basin. Based on these data, for the given parameters of effective porosity 5%, rock specific gravity 2.5, oil density 0.8, a oil saturation of 1.3% is calculated for source rocks of most of oil fields in China; and 6.4% for the optimum source, rocks in Daqing field, 0.4% for those in Han No. 1 well. It is quite evident that those fields all have a very low oil saturation, much lower than the oil saturation threshold (20% -30%) for source rocks applied by Ozkaya.

Some conclude that primary migration is processed in a mode of oil water miscible phase. But the author considers it a kind of subjective surmise. A dewatering experiment of crude oil under high pressure and temperature( $>10^6$ Pa, $>420$ K) shows that oil and water have a separative rather than a miscible property on the basis of which the dewatering of crude oil and the disintegration of already-formed emulsoids are processed. Moreover, the boiling point of water is low relating to the major hydrocarbons in the oil. Even if the water moleculars dewatered from clay minerals will evaporate and escape; and it is difficult for those hydrocarbons generated in a source rock during peak generation to form emulsoids with water. Even the formation does occur, they are unstable because the properties of crude oil will change along with the variation of environmental temperature as soon as their separation from the source rock. For example, the crude oil from California has a dynamic viscosity of 50 mPa · s at 100 ℃; and of 1200 m Pa · s when the temperature decreases to 50 ℃; But for pure water at 100 ℃, the dynamic viscosity is 0.28, and 0.55 at 50 ℃ (from P. Ungerer et al). Because of the very variation resulted in the property differentiation, the disintegration of oil-water emulsoids occurs.

Furthermore, a distinct boundary between the petroliferous section and non-petroliferues section can be been based on the observations of oil-bearing sandstone. In fact, this kind of oil-water separation has refuted the theory of oil-water emulsoids within a stratum.

## 4 The Pyrolysates and Petroleum Bear a Poor Correlation Constituently

Constituently there is very much difference between the pyrolysates of Kerogen (or mudstone) and crude oil/natural gas. Their poor correlation is one of the major drawbacks in the theory of oil generation from kerogen thermal degradation, and its major expressions are as follows:

### 4.1 Dryness or wetness of gaseous hydrocarbon

Based on many data of thermal simulation the gas generated from the thermal degradation has a low content of methane, a low ratio of $C_1/C_2^+$, and a high degree of wetness. On the contrary, in natural gas (including the associated gas of petroleum), the content of methane and $C_1/C_2^+$ ratio are higher, but its wetness is low. This difference is a well-known fact, needless to go in details.

Some consider that the causes for the difference are due to "the differentiation of light-heavy during the process of natural gas migrated from source rock into the reservoirs" (Li Zhi, 1990). However the author thinks that there may be some kind of priorities during the migration of gaseous hydrocarbons from source rock to reservoir, but it is only a migrating sequence, just as the differentiation of retention time in GC analysis, because the difference of this retention time and that of various components is very brief relating to the time necessary for the formation of a commercial oil/gas pool. Moreover, the process of the gas of the pool from the source rock into the reservoir through the carries bed is neither like a mono-injection in chromatography, nor in the way that methane formed early migrates first, and afterwards ethane formed later in gas source rock migrates. Therefore, the light-heavy differentiation similar to that in chromatography will not occur in the formation of commercial gas pool. In geological bodies, natural gas pool mainly consisted of ethane or propane has not been found, which is a strong evidence. In addition, it is difficult to explain the relation between the two isomers of butane in terms of differentiation: the boiling point of $n$-butane is 12 ℃ higher, and the absolute entropy $S^0$ is 3.7 cal. (deg · mol)$^{-1}$ higher than $i$-butane and its retention time in chromatographic analysis is longer. If the primary migration has chromatographic affection, there must be many heavy components remained in rocks, and that more $i$-butanes than $n$-butane in natural gas, but in fact $n$-butane is generally higher than $i$-butane in natural gas, and vice versa for absorbent hydrocarbon in a rock. Obviously, no differentiation occurs in primary migration.

### 4.2 Difference in the content of *n*-butane and *i*-butane in gaseous hydrocarbon

For a chemical reaction, the content relation between isomers of the products from the reaction is a very important index to distinguish the reaction and evidence to study its characteristics and conditions. The butane isomers in the hydrocarbon gas of petroleum are those simplest isomers with lowest carbon number among the saturated hydrocarbon compounds. Undoubtedly they are objects of studying the chemical reactions for petroleum generation.

Based on the data in table 1, much difference is found in the contents of $i$-butane ($iC_4$,) and $n$-butane ($nC_4$) in the hydrocarbon gas of petroleum and gas of kerogen pyrolysis; and also in $iC_4/nC_4$ of hydrocarbon gas of petroleum and absorbent hydrocarbon escaped from the rock (source rock). Evidently that the hydrocarbon gas is like neither the pyrolytic gas of kerogen thermal simulation nor the absorbent gas in source rock.

According to A. T. James (1984), the most influential factors over $iC_4/nC_4$ are the degrees of biodegradation. However biodegradation is in a trend that the ratio increases as the degrees of the reaction are enhanced. Thus geological factors will not influence the reliability of the correlation in table 1.

**Table 1. Data of $iC_4/nC_4$ for various types of samples**

| Sample Type | $iC_4/nC_4$ | Notes for Samples |
|---|---|---|
| Oil soluble gas of crude oil in high pressured layer | 0.309 – 0.355 | Based on the alatistics of 40 wells in Eogene, Jiangsu, China (gas wholly secreted from the crude oil of high pressured layer). |

续表

| Sample Type | $iC_4/nC_4$ | Notes for Samples |
|---|---|---|
| Hydrocarbon gas in petroleum | 0.289 | Selected from samples above no biodegradation, no pressure effects and no difference of preservation for these samples. |
| Rock absorbent gas | 1.56 | The statistic values from four wells in an oil field in China. The depth is from 1500 m ($R_0$ = 0.5) to 4000 m ($R_0$ = 1.23). |
| Pyrolytic gas of kerogen: | | |
| Major stage of hydrocarbon generation | 0.6 – 1.4 | Gas processed on the basis of thermal simulation of Lizhi et. al. |
| Early stage of hydrocarbon generation | 1.2 – 2.0 | |
| Late stage of hydrocarbon generation | 1.2 – 2.0 | |

## 4.3 Problem of alkene

Any thermal simulation (except for there is water condition) will result in a series of abundant alkenes; while any crude oil or natural gas has very low or undetectable amount of alkenes. They are acknowledged facts, and needless to say they constitute a controversy for the theory of oil generation from thermal degradation of kerogen. It will not be solved that there is no alkene in the products of thermal simulation under aqueous condition because almost all the free water in rock was lost early at the time when the sediments had been buried deep into the threshold of oil generation. It is difficult for the amount of water released from heated clay minerals to form the medium condition for an aqueous pyrolysis.

## 4.4 About other components

Pyrolytic products such as from the thermal simulations of kerogen and asphaltene (including asphaltene of crude oil and rock extracts) contains multiple series of compounds which are not found in crude oil or there is much difference in content between crude oil and pyrolytic products. For example the content of aromatic compounds with no substituents in aromatic nucleus is higher in pyrolytic products than in crude oil, such as the case for benzene, naphthalene and anthracene. So far as the correlation with benzene is conceded, the concentration ratio of crude oil and pyrolytic products is over 10. The relative content of benzene is particalurly high in the pyrolytic products in type kerogen, the concentration for the two is much higher.

Some compounds such as phenol, methyl phenol and methyl ethyl phenol in pyrolytic products has a content far more higher than in crude oil. These compounds are not detected in many crude oils and petroleum gases.

Moreover, the total amount at aromatic compounds holds a very low content in non-marine crude oil for example in the crude oil of Daqing, China, it is about 10% ( W/W) , however, the content of kerogen pyrolytic products from non marine layers can be as high as 20% , and will be higher in type kerogen, such instances will not discussed in detail.

## 5 Pool Reliability on Oil/Source Rock Correlation

With abundant data of oil/source rock correlation, the author has studied the source rocks of crude oil from No. 2 well in the Tarim Basin, Jiangsu oilfield, east China Sea and Han No. 1 well. The results show that the correlation of crude oil with the source rock defined by the theory of oil generation from kerogen is poor in reliability. Because of the limit space, the paper will illustrate it with only two examples.

In view of all the analytical data of geochemistry taken from Han No. 2 well in Chaidam Basin, we found that the maximum organic carbon of the 34 samples collected at the well depth of 195 m to 6018 m is 0. 23% , while for those of maturation from $N_2 - N_1$ ( well depth 3350 – 5454 m) , the mean organic carbon is only 0. 17% , the average value of hydrocarbon content is 66 ppm, and the mean saturated hydrocarbon is 32. 4%. These parameters indicate that all the "source rocks" being analysed from Han No. 2 can not qualify the commonly recognized concepts of being a source rock. Never the less, crude oil was met at the depth of 5475 – 5522 m in Han No. 2 well. The content of saturated hydrocarbons reaches 64. 75% after many light hydrocarbons have been lost, among which there is certain amount of paraffin hydrocarbons higher than $C_{22}$, and evidently, their content are more than that of *n*-alkanes with higher carbon number in rock extracts. These *n*-alkanes have no source to trace and no rock to correlate with. Thus it can be seen that such kind of oil/source rock correlation and the source rock itself may not be counted as reliable.

Now the non-correlativity of oil/source rock will be discussed with a case study of oil/source rock correlation from a petroliferous structural in an oilfield in China, in terms of bio-markers.

Figure 1 is a correlation of 217 m/e ionic chromatograms in GC/MS analyses of rock extracts and saturated hydrocarbons. In the figure, the regular sterane distributive areas in No. 843463,843464 and 84346 of 217 m/e ionic chromatograms being derived from 13 source rocks, represent the source rocks in oil/source rock correlations; While the oil (3792 m) is similar regular sterane chromatogram representing 7 crude oil samples, they are produced from the intervals in which the 13 source rocks distributed, at the depth 2300 –4100 m, and their 217 m/e ionic chromatograms are similar, the oil sample (3792 m) can be a representative of "oil", the representative compounds will be shown in the determinative by peak number table (table 2).

From the correlation in Figure 1, it can be seen that the correlation between

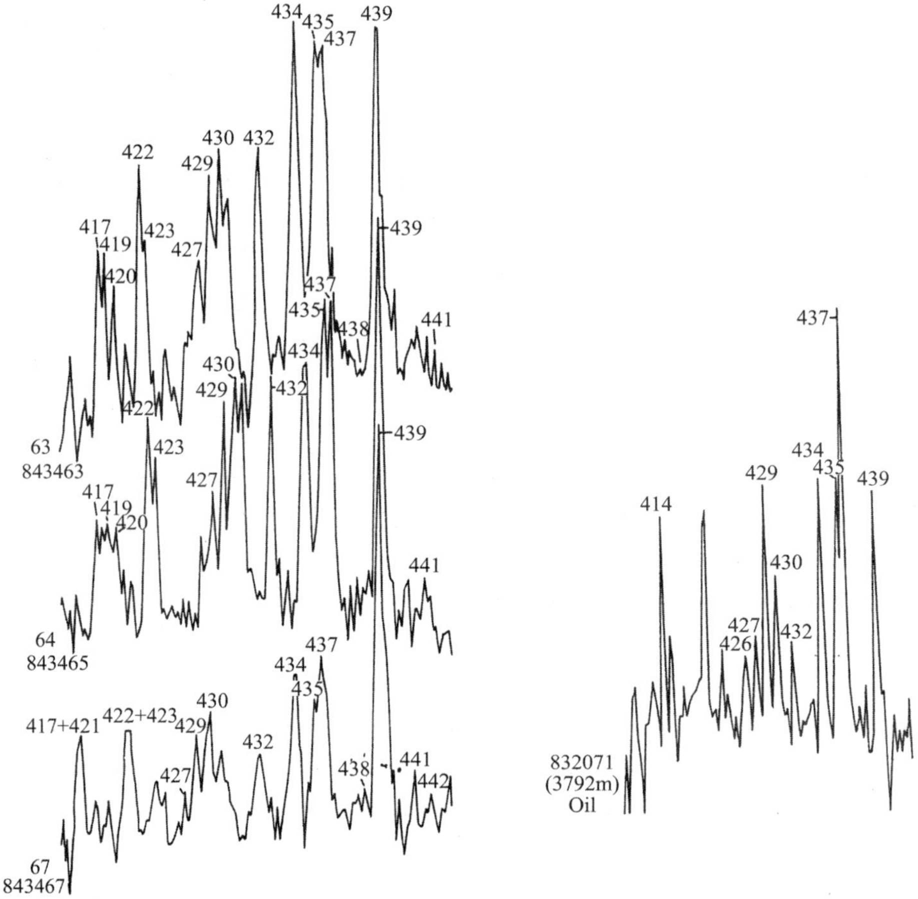


**Fig. 1. The correlation of 217 m/e ionic chromatograms in Gc/Ms analyses of rock extracts and saturated hydrocarbons**

source rock and oil is impossible for the 12 regular (normal) steranes ($C_{27}$ – $C_{29}$) with different patterns, and there is no generality and similarity between the two.

**Table 2. Determinative table**

| Peak No. | Molecular weight | Molecular formula | Name of compound |
|---|---|---|---|
| 417 | 372 | $C_{27}H_{48}$ | 5α(H),14α(H),17α(H) –20S – cholestane |
| 419 | 372 | $C_{27}H_{48}$ | 5α(H),14β(H),17β(H) –20R – cholestane |
| 420 | 372 | $C_{27}H_{48}$ | 5α(H),14β(H),17β(H) –20S – cholestane |
| 421 | 400 | $C_{29}H_{52}$ | 13β(H),17α(H), –20R –24 – ethyl rearrangement cholestane |
| 422 | 372 | $C_{27}H_{48}$ | 5α(H),14α(H),17α(H) –20R – cholestane |
| 423 | 400 | $C_{29}H_{52}$ | 13α(H),17β(H), –20S –24 – ethyl rearrangement cholestane |
| 427 | 386 | $C_{28}H_{50}$ | 5α(H),14α(H),17α(H) –20S –24 – methyl cholestane |
| 429 | 386 | $C_{28}H_{50}$ | 5α(H),14β(H),17β(H) –20R –24 – methyl cholestane |
| 430 | 386 | $C_{28}H_{50}$ | 5α(H),14β(H),17β(H) –20S –24 – methyl cholestane |
| 432 | 386 | $C_{28}H_{50}$ | 5α(H),14α(H),17α(H) –20R –24 – methyl cholestane |
| 434 | 400 | $C_{29}H_{52}$ | 5α(H),14α(H),17α(H) –20S –24 – ethyl cholestane |
| 435 | 400 | $C_{29}H_{52}$ | 5α(H),14β(H),17β(H) –20R –24 – ethyl cholestane |
| 437 | 400 | $C_{29}H_{52}$ | 5α(H),14β(H),17β(H) –20S –24 – ethyl cholestane |
| 439 | 400 | $C_{29}H_{52}$ | 5α(H),14α(H),17α(H) –20R –24 – ethyl cholestane |
| 441 | 414 | $C_{30}H_{54}$ | 4 – methyl –24 – ethyl cholestane |

## 6 Conclusion

The paper discusses the various drawbacks in the theory of oil generation by the thermal degradation in light of four aspects i. e. "immature" oil, primary migration of oil and gas, pyrolytic products and oil/source rock correlation. The summary in each section indicates that all the commercial oils (except for condensate oil) may not be formed by the thermal degradation and the theory does not conform to the practice in geology and geochemistry in numerous respects. The author considers that the debate on the origin of petroleum has already lasted for more than a century. One of obstacles in it is that people always regard fine grained sediments (mudstone) as major source rocks, which mislead the genetic study of petroleum into a twisted and tedious wrong way. However, the theory of oil generation with synthetic effects in a waterlogged and closed system will probably provide up a new way for the final solution of petroleum genetic theory.

## References

[1] Баженова OK. Образование Нефть на Неьольших Глубинах[J]. Геолоzuя Нефmь u Газа, 1990(7):2-7.

[2] MacKenzie AS, Maxwell JR. Organic maturation studies and fossil fuel exploration [J]. 1981:239-254.

[3] He Zhigao. Experimental Petroleum Geology, 1982, 4(1):1-9.

[4] Özkaya I. Computer simulation of primary oil migration in Kuwait[J]. Journal of Petroleum Geology 1991,14(1):37-48.

[5] Проскуряков BA. Химия Нефть и Газа, Извдателвство. Хuмuя Ленинград, 1981.

[6] Li Zhi, Zhu Meiqian, Mao Hongyi. Oil & Gas Geology, 1990, 11(3):247-259.

[7] James AT, Burns BJ. Microbial alteration of subsurface natural gas accumulations[J]. AAPG Bull, 68(8):957-960.

[8] He Zhigao. Experimental Petroleum Geology, 1981,3(3): 197-202.

[9] Huang Difan. The evolution of non-marine organics and the mechanism of hydrocarbon generation[M]. Beijing: Petroleum Industry Press House, 1984.

[10] He Zhigao. A model of oil-generation in a waterlogged and closed system[J]. J Southeast Asian Earth Sci, 1991, 5(1-4):456-469.

# 9. 煤不成油的理论基础 假石油与真煤焦油

——再议煤不成油

# Coal does not generate petroleum—on the discussion of the theoretical basis

何志高

## 1 讨论前提

为了严肃学术上的争论,有必要首先指出,笔者所讨论的“煤不成油”中的“油”历来是指石油,是指工业化中主要的工业用石油(或商业石油),而不是氯仿沥青、沥青脉,更不是煤焦油;“煤”是指广泛使用的固体可燃矿产,而不是煤系地层。这种定义及概念是与世界范围100多年来所争论的石油成因理论一脉相承的。本文也顺便议及煤系地层成油。

## 2 煤不能成油的理论基础

笔者认为,除了提出证明煤不成油的各种方法以外,还需要从理论上进行必要的讨论。鉴于这个栏目篇幅有限,只能是一种简议。而且,这种理论上的深入研究,正是年富力强的学者们义不容辞的责任。何况,半个世纪以来,这种研究资料已相当丰富,*Fuel*杂志是20世纪30年代初创刊的,到50年代就刊载了关于煤的结构、组成、性能等大量的用近代分析测试技术所提供的研究成果。正是这些研究成果,才有可能为后人60年代建立干酪根晚期热降解成油理论打下基础;也正是这些研究成果,才使煤成油理论虽然最早由杰出的俄国著名学者罗蒙诺索夫创立,直到今天仍然没有广泛的支持者,后继乏人,地质实际不支持,理论上也缺乏依据。

笔者综合前人与本人的研究成果,认为煤成油在理论上是无基础的、荒谬的,其依据是:

(1)煤的氢碳原子比最低。对比煤与石油、油页岩中有机质、分散有机物质(干酪根)等,煤的氢碳原子比是最低的,比干酪根成油学说中所确定的生油能力最低的Ⅲ型干酪根的氢碳原子比还要低。从这个意义上而言,煤的生油能力是不值得考虑的。

(2)在煤的分子结构中,氢原子大多数易与氧结合而生成水。构成煤的原始物质,例如高等植物主要是纤维素。半纤维素和木质素,它们在理论上都归属在碳水化合物一类,这类化合物形成它们的反应物就是碳(二氧化碳)和水作用的光合作用,它们形成后在没有水的还原环境受地热的作用,必定是失水留碳,而不可能碳与氢结合形成石油烃类。当然,从元素的电负性上进行分析和从热力学理论分析均可得出同样的结论。

(3)煤的芳构化,即随变质程度增加,煤的分子结构向芳香环结构变化是由很多方法检验、验证并早已形成定论的事实。特别是红外光谱中反映芳碳环结构的共轭大兀电子的随兀电子的离域作用增强,而使伸缩振动能量减弱,1600 $cm^{-1}$谱带向低波数位移的事实也早已由笔者等对7种煤进行研究后所确立。

芳构化必然使得碳原子趋向稳定,从根本上否定了煤形成大量的相对分子较小的石油烃类化合物的可能。

芳构化后为数不多的侧链,一则因数量有限,再则断裂后也更易于环化,加之较丰富的初级氢原子也更易与剩余的氧原子结合形成水,因此,侧链形成石油烃类的作用也是很弱的。

(4)再探讨煤成油提前的可能性。有人认为,煤成油的作用可以发生在成煤早期。笔者认为,这种观点也是缺乏理论根据的。前已述及,煤的原始物质,如木质素、纤维素、半纤维素均是不溶于水的。没有水就没有水解,也没有微生物的作用,也就没有相应的生化作用,没有非热分解的化学反应。在最初开始沉积的阶段,表面及空隙虽有水,但是与水的接触面(表面)、接触时间(空隙)少而短暂,在厌氧的还原环境下,只可能形成甲烷。由于随同沉积的碳酸盐物质较少,形成甲烷的生化反应所产生的二氧化碳无法与碳酸盐反应形成碱性环境的重碳酸盐,使水体始终处于 $pH < 7$ 的酸性介质中,而无形成比甲烷更大分子的石油烃类的可能。

综上所述,从成煤的原始有机物质到沉积、成岩直至石墨化的整个过程,都不可能形成石油。至于有人提出的开采煤矿时有油冒出来使得矿工打湿裤子的油量,在工业用油的滚滚滔滔的油流之中,在亿吨级煤炭储量之中都顶多称得上杯水车薪。何况,那很可能是煤焦油而非真正的石油。笔者建议,取其油样对其进行分析测试,以辨其真伪。

## 3　笔者讲一个与此有关的故事

记得30多年前的1968年,笔者被派去山西临汾县检查“油苗”。当时,县工业局以一处山谷水溪漂出大量浮油为事实,向地质部报告请求勘查。地质部军代表批文咸阳第三普查大队检验。第二天,笔者和李炳刚出发,当日到达临汾,次日由工业局主管陪同,从县城步行约15千米到达出油处。经现场勘察,沿着出油水沟往上追踪50多米,一个由石块砌成的土坝构成一个大的烧结焦炭的坑,从数十米远的山谷山坡边的煤窑中挖出来的煤炭就在这个大坑中被烧成焦炭,而水就是通过烧炭坑的底部,带着煤焦油涌出地面,然后越积越多,累计有一吨多。采回的样品及其分析测试与现场观察结论完全一致——这是假石油,真煤焦油。与我们同时(晚我们一天)到达的还有石油工业部指令的银川石油勘探局的杜棚等两位同仁,他们得出的结论完全与我们的一致。30多年过去了,再也没有群众来报矿的,也没有听到对这个结论表示质疑的。

笔者还处理过安徽、湖南等地的煤焦油误作石油的真假石油的鉴定事件。因此,煤矿中遇到油,除了确实存在的断层可能引入远距离的石油之外,多数是煤焦油。其实,如果你承认煤的自燃现象存在,那么,就不难理解煤矿中涌出煤焦油来的事实。

## 4　关于煤系地层成油

笔者认为,根据湿封闭体系综合效应成油论,凡是具备湿封闭体系条件的任何沉积地层都可以有生成石油的基本条件,当然,煤系地层也会发育生成石油的基本条件。笔者记得30年前一位地质同仁姚志温先生告诉笔者,苏联有位学者讲煤与石油的关系是同心圆的关系。笔者认为,这种同心圆的关系是可能存在的,但不是必然的,不是普遍存在的。

本文原载于《石油知识》总第92期(2002)

# 10. Mudrock not being oil-gas source rock

He Zhigao, Shen Liduo

(The Central Laboratory of Petroleum Geology, Ministry of Geology and Mineral Resources, Wuxi, Jiangsu, P. R. China, 214151)

**Abstract**: This paper reports the ratio for isobutane and normal-butane ($iC_4/nC_4$) of hydrocarbon gases released from 102 samples in closed pulverized apparatus, which come from seven wells in a oil-gas basin, China, and other relative data. By discussing this phenomenon in the term of absorption theory, we find that these ratio of samples from these seven wells (from ca. 1500 to 4000 m) don't rechieve the range of this ratio of oil-soluble and associated gases, from which it can be concluded that the mudrock samples to be studied cann't be oil-gas source rock. Oil-soluble and associated gases are generated by the bio-decarboxytation of aliphatic acids in waterlogged and closed system instead of by the pyrolysis of organic matter in mudrock.

**Key words**: mudrock; butane absorption; aliphatic acid; bio-decarboxylation

## Introduction

There are many advantages to study the relative abundance of butane isomers:

(1) Using simple analytic technique we can obtain a lot of data which not be affected by pollution and the accurate quantitative data can be used.

(2) It is not necessary to deal with the primary and secondary migration of hydrocarbons, which is difficult in study of petroleum geochemistry.

(3) Butane isomers have stable and similar chemical property.

(4) In petroleum hydrocarbons, butane possesses the simplest isomer component (only two).

(5) Butane isomers present in both oil phase and gaseous phase. Although the

complex construct compounds in petroleum, such as strane, have been studied, the meaning and conclusion are obvious to choose butane isomers as subject of study.

## 1 The distribution of ratio of butane isomers in mudrock

Table 1 gives us the $iC_4/nC_4$ ratios calculated from hydrocarbon gases which release from mudrock and coal which samples from seven wells in East Sea basin, China, by pulverizing in sealed apparatus. The data of corresponding depth of oil formation and the vitrinite reflectivity $R_0$ of PF-1 well are also given in Table 1.

**Table 1. The ratios of $iC_4/nC_4$ in mudrock samples from seven wells**

| LHT – 1 | | FG – 1 | | PF – 4 | | QY – 1 | | ZY – 1 | | BYT – 2 | | TJ – 1 | | Depth of oil-formation (m) | $R_0$ (coal) |
|---|---|---|---|---|---|---|---|---|---|---|---|---|---|---|---|
| Depth (m) | $iC_4/nC_4$ | Depth (m) | $iC_4/nC_4$ | Depth (m) | $iC_4/nC_4$ | Depth (m) | $iC_4/nC_4$ | Depth (m) | $iC_4/nC_4$ | Depth (m) | $iC_4/nC_4$ | Depth (m) | $iC_4/nC_4$ | | |
| 1507 | 1.37 | 1406 | 1.12 | 1810 | 1.33 | 1850 | 1.73 | 1466 | 1.23 | 2216 | 1.55 | 1600 | 2.98 | | |
| 2090 | 2.86 | 1496 | 1.25 | 2298 | 1.23 | 2080 | 1.31 | 1772 | 0.95 | 2415 | 1.27 | 2321 | 1.57 | 2309 | |
| 2128 | 1.58 | 1645 | 1.15 | 2568 | 1.48 | 2290 | 1.15 | 1953 | 0.59 | 2663 | 0.65 | 2590 | 3.28 | | |
| 2283 | 1.56 | 1814 | 2.64 | 2773 | 1.68 | 2321 | 3. 20 | 2155 | 1.10 | 2775 | 2.64 | 2760 | 1.61 | 2972 – 2975 | 0.67 (2864 m) |
| 2650 | 1.41 | 2000 | 2.97 | 2954 | 1.90 | 2324 | 1.80 | 2277 | 0.83 | 3038 | 1.52 | 2864 | 1.73 | | |
| 2993 | 1.01 | 2205 | 1.79 | 3000 | 3.10 | 2559 | 1.81 | 2582 | 1.96 | 3276 | 0.73 | 2870 | 2.13 | 2981 – 2986 | |
| 3159 | 1.76 | 2341 | 1.32 | 3002 | 2.21 | 2838 | 1.23 | 2813 | 1.46 | 3401 | 1.88 | 2960 | 0.97 | | |
| 3325 | 2.48 | 2405 | 2.09 | 3090 | 0.88 | 3084 | 0.82 | 3045 | 0.94 | 3404 | 2.47 | 3045 | 1.84 | | |
| 3577 | 3.53 | 2608 | 2.00 | 3165 | 0.91 | 3174 | 1.04 | 3075 | 1.55 | 3447 | 0.98 | 3268 | 2.02 | 3128 – 3792 | 0.82 (3579 m) |
| 3636 | 3.10 | 2678 | 1.30 | 3545 | 2.21 | 3365 | 1.60 | 3151 | 2.35 | 3560 | 2.27 | 3381 | 2.11 | | |
| 3637 | 2.87 | 2799 | 0. 84 | 3645 | 2.31 | 3400 | 1.87 | 3227 | 1.28 | 3629 | 1.53 | 3579 | 2.14 | | |
| 3712 | 3.27 | 2880 | 1.42 | 3655 | 2.00 | 3537 | 1.79 | 3530 | 1.59 | 3757 | 1.01 | 3634 | 1.86 | | |
| 3737 | 3.42 | 2934 | 1.35 | 3750 | 1.46 | 3688 | 1.34 | 3612 | 0. 96 | 3784 | 1.14 | 3709 | 1.86 | 3777 – 3792 – 3806 | 0.86 (3870 m) |
| 3770 | 2.50 | 3133 | 1.18 | | | 3870 | 1.87 | | | | | 3773 | 1.46 | | |
| | | 3378 | 1.35 | | | 4071 | 1.44 | | | | | 3870 | 1.72 | | |
| | | 3651 | 1.19 | | | 4140 | 1.86 | | | | | | | 4267 – 4271 | |
| | | 3930 | 2.20 | | | | | | | | | | | | |
| | | 4072 | 2.12 | | | | | | | | | | | | |

As we see from Table 1, the average of 102 samples (the most are mudrock) is 1.73, ranging from 0.59 to 3.53. The percent of samples, of which the $iC_4/nC_4$ ratio more than 0.6, is about 99%. There is no obvious changeable tendence between the depth and this ratio for single well. But the ratios for coal samples in two wells decrease with the increase of depth, which from 3.5 to ca. 1.5.

## 2 Discussion

Many papers regarding the distribution of butane isomers have been published. We don't repeat here. But these data can be summed up in following: (1) The $iC_4/nC_4$ ratio of oil-soluble and associated gases, which not be degraded by microorganism, are mostly less than 0.6, generally 0.3, but if degraded, this ratio is higher, even up to 9; (2) For absorbed gas of mudrock, this ratio is mostly more than 0.6; (3) The ratio of hydrocarbon gases generated by anaerobe fermentation is about 0.6.

According to the ratio of $iC_4/nC_4$, the samples can be arranged from high to low in following way:

| | |
|---|---|
| Absorbed gas of coal | 2.32 (average of 6 samples) |
| Absorbed gas of mudrock | 1.5 |
| Gas from anaerobe fermentation | 0.6 |
| Oil soluble and associated gases | 0.2 – 0.6 |

As we known, the absorption of material can be classified as two categories, chemical and physical, their absorb forced are chemical bonding and Van der Waals force separately. With regard to $i$-butane and $n$-butane, their chemical force are almost same, but for Van der Waals force, former is less than later. In this situation, we can suppose that the concentrations of $iC_4$ and $nC_4$ are same, the surface of solid (such as mudrock and coal) absorbs more $nC_4$ than $iC_4$, which will result in $iC_4/nC_4$ ratio less than 1. In other words, if oil-soluble and associated gases were generated by the pyrolysis of micromolecule organic matter in mudrock, the $iC_4/nC_4$ ratio of oil-soluble and associated gases ought to be similar to that of absorbed gases of mudrock from which oil-soluble and associated gases originate. However, we can see as the data given in Table 1 and about oil-soluble and associated gases that there is not any corresponding relationship between them. This fact indicates that oil-soluble and associated gases don't come from the pyrolysis of organic matter in mudrock. At this time, we can further conclude that crude oil doesn't originate from the pyrolysis of organic matter in mudrock.

We believe that hydrocarbon gases are generated by the bio-decarboxylation of fatty acid in waterlogged and closed system. This reaction occurs in the system saturated by water, such as sandstone. The pentanoic acid has four isomers, but only three present in nature, they are *n*-pentanoic acid, 2-methyl butanoic acid and 3-methyl butanoic acid. The first two decarboxylate to form *n*-butane and the last to form *i*-butane. The abundance of this three acid isomers are same in specific system. Therefore, the $iC_4/nC_4$ ratio of nature gas generated by decarboxylation of pentanoic acid is approximate 0.5 depending on quantity of each isomer of pentanoic acid and not considering the difference of decarboxylation reaction of this three isomers. This value fits with that of oil-soluble and associated gases unaffected by biodegradation. According to the study of author (He Zhigao, 1986), with the increase of fermentation time, the biodegradation of long chain *n*-pentanoic acid will increase and the concentration of it in solution gradually decrease, which will result in the further deduce of $iC_4/nC_4$ ratio. So that this ratio for most oil-soluble and associated gases is less than 0.5, even 0.2.

We have normalized the $C_1 - C_4$ oil-soluble gases and $C_2 - C_5$ fatty acids separately. The left is distribution curve of components of oil-soluble hydrocarbon gases and light is the distribution curve of $C_2 - C_5$ aliphatic acids. Obviously, these two curves are very similar. This evidences the view of the formation of hydrocarbon gases by bio-decarboxylation of fatty acids in water.

## 3 Conclusion

As mention above, the oil-soluble and associated hydrocarbon gases are not produced by pyrolysis of organic matter in mudrock instead of by bio-decarboxylation of aliphtic acids in waterlogged and closed system. This means that the mudrock in oil-gas basin is not oil-gas source rock.

## References

[1] He Zhigao. A model of oil-generation in a waterlogged and closed system[J]. J Southeast Asian Earth Sci,1991, 5:465 -469.

[2] He Zhigao. On the origin of petroleum hydrocarbon gases[J]. Acta, Geologica Sinica, 1986,60(4):57 -65.

[3] He Zhigao. Discussion on the eevolution of hydrocarbons in the light of the theory of thermodynamics[J]. Experimental Petroleum Geology,1982,4(1):1 -9(in Chinese).

# 11. Generation of petroleum hydrocarbon gases directly by microbial fermentation

He Zhigao, Shen Liduo

(The Central Laboratory of Petroleum Geology, Ministry of Geology and Mineral Resources, Wuxi, Jiangsu, P. R. China, 214151)

**Abstract**: This paper mainly reports the experimental method that microbial fermentation directly generates petroleum hydrocarbon gases and the results which is useful for understanding the mechanism of petroleum hydrocarbon gases and strongly support the theory that petroleum is formed under waterlogged and closed system.

**Key words**: microorganism; fermentation; petroleum hydrocarbon gases; generation

## 1 Experimental method of microorganism fermentation

### 1.1 Microorganism

The anaerobic microorganism group from marsh gas fermentation is used in experimental, which consists of methane bacteria ( $> 4.5 \times 10^9$ colony unit/milliliter), hydrogen-HAc (ethnoic acid) bacteria ( $>10^{10}$ ), ethnoic acid bacteria ( $>10^5$ ) and so on, such as methanobacterium, methanococus and methanospirillum.

### 1.2 Nutrient

Molasses from sugar factory is used as nutrient in this experimental. Its content of sugar is over 40%.

### 1.3 Experimental apparatus for fermentation

A 250 mL suction flask is placed in heater case and some conducts are connected to flask to collect gases from fermentation.

### 1.4 Procedure

100 mL molasses(10%) and 100 mL marsh fermentative solution(10%) are put into suction flask separately. Nitrogen gas is used to wash the space of flask to drive air out and then the flask is sealed to prevent from air. The temperature of fermentative reaction is 45 ℃. During reaction, the gases are collected and analysed in certain interval of time. Whole experiment takes 40 days.

## 2 Experimental Results

The amount of gases-generating of molasses is about 86 mL/g. Total 860 mL gases is generated in whole experiment, which mostly is produced in the first four days. When reaction achieves the high peak of generating gasses, about 300 mL/d gases are generated. The analytic results of gases are listed in Table 1.

**Table 1. The results of composition analysis of hydrocarbon gases**

| Sampling No. | Reaction time | Composition of hydrocarbon gases(%) | | | | | | | | $CO_2$(%) |
|---|---|---|---|---|---|---|---|---|---|---|
| | | $CH_4$ | $C_2H_6$ | $C_3H_8$ | $iC_4H_{10}$ | $nC_4H_{10}$ | $iC_5H_{12}$ | $nC_5H_{12}$ | $C_6^+$ | |
| 1 | 6 h | 98.09 | 1.39 | 0.07 | 0.17 | 0.08 | 0.14 | 0.05 | * | 29.21 |
| 2 | 12 h | 92.93 | 3.66 | 0.40 | 0.22 | 0.99 | 0.52 | 0.87 | 0.40 | 58.59 |
| 3 | 29 h | 92.47 | 3.19 | 0.5 | 0.42 | 1.63 | 0.74 | 1.05 | * | 79.7 |
| 4 | 36 h | 96.50 | 0.4 | * | * | 1.40 | 0.68 | 1.01 | * | 78.43 |
| 5 | 95 h | 92.54 | 0.13 | 0.38 | 1.79 | 1.38 | 2.07 | 0.37 | 1.31 | 68.65 |
| Average | | 94.51 | 1.75 | 0.27 | 0.52 | 1.10 | 0.83 | 0.67 | 0.34 | 62.92 |

Note: * Undetected

## 3 Discussion

### 3.1 The characteristic of composition of petroleum hydrocarbon gases

As well known, the composition characteristic of petroleum hydrocarbon gases distinctively differ from that of gases from other origin and field. From Table 1, it can be seen that the composition of gases generated in microbial fermentation has the same characteristic as petroleum associated natural gases:

(1) The content of heavier hydrocarbon gases (molecular weight more than or equal to ethane) is over 5%.

(2) There is a significant content of $C_6^+$ hydrocarbons (average 0.34%).

(3) $nC_4/iC_4$ ratio over 2, which fits in with the content and change tendence of fatty acids in marsh fermentation liquid.

(4) The relative distribution of $C_1 - C_6^+$ hydrocarbon gases is agreeable to that of petroleum associated gases, which is also in line with the relationship of content of fatty acids and the regulation of decarboxylation.

### 3.2 About carbon dioxide

For most oil fields in world, the content of carbon dioxide in oil-soluble gases and associated gases is lower. But it can be seen in Table 1 that the content of carbon dioxide is higher. It seems to be contrary to "the composition of real petroleum gases". According to the study of author, a waterlogged and closed system is saturated by water and there are carbonate rocks and cements in dribs sediments. Carbon dioxide must take part in following reaction so that its content will decrease and even disappear:

$$CO_2 + CaCO_3 + H_2O = Ca(HCO_3)_2$$

Although we can't find out the real reason for too higher content of carbon dioxide in gases produced by this experiment, comparing with common marsh gases, this phenomenon does not affect the conclusion of this paper.

## 4 Conclusion

From experimental results, hydrocarbon gases generated by microorganism fermentation contain a significant $C_2^+$ hydrocarbon gases, especially $C_6^+$ which only present in oil. The ratio $nC_4/iC_4$ completely agrees with that of petroleum associated gases instead of rock adsorbed gases. All these phenomenon indicate that petroleum hydrocarbon gases come from the decarboxylation of fatty acids solving in water by microorganism action.

(Note: $C_6^+$—all hydrocarbons including 6 carbons and more; $C_2^+$—all hydrocarbons including 2 carbons and more.)

**Acknowledgements**

We acknowledge Mr. Zhang Chunying, a senior engineer in Academy of Exploration and Exploitation of Daqing Oilfield, China, for arranging experiment and analysis. We also thank professor Qian Zeshu (Zhejiang Agriculture University) for his guiding and Mr. Lin Xineng (Wuxi Second Pharmaceutical Factory) for his helping.

**References**

[1] He Zhigao. On the origin of petroleum hydrocarbon gases[J]. Acta Geologica Sinica, 1986,60(4):57-65.

[2] He Zhigao. A model of oil-generation in a waterlogged and closed system[J]. J Southeast Asian Earth Sci,1991,5(1-4):465-469.

[3] Qian Zeshu, Ming Hang. Marsh fermentation microbiology[M]. Zhejiang Hangzhou: Zhejiang Science and Technology Publishing House, 1986 (Chinese).

# 12. 微生物发酵实验生成真正石油烃类气的证据

# Generation of petroleum hydrocarbon gases directly by microbial fermentation

何志高,沈力铎

(地质矿产部石油地质中心实验室,江苏无锡,214151)

**摘要:** 本文报道由微生物发酵实验直接生成真正石油烃类气的实验方法及其资料,证据确凿。争论了一个多世纪的石油成因理论由此而取得了新的突破。这种证据有力地验证了笔者早先创立的湿封闭体系综合效应成油理论。

**关键词:** 微生物;发酵;石油烃类气;生成

## 1 微生物发酵实验方法

### 1.1 微生物

沼气发酵中的厌氧微生物群,主要有产甲烷菌(浓度大于 $4.5\times10^9$ 菌落单位/mL),如甲烷杆菌(methanobacterium)、甲烷球菌(methanococus)、甲烷螺菌(methanospirillum)等以及产氢产乙酸菌($>10^{10}$)和产乙酸菌($>10^5$)等。

### 1.2 营养源

糖蜜,由糖厂出售,含糖量约为40%。

### 1.3 发酵实验装置

发酵反应瓶为250 mL吸滤瓶,加上收集气体的管道和量气瓶,安装在加热恒温箱中。

### 1.4 操作方法

将10%的糖蜜100 mL,放入吸滤瓶,再加入20%的沼气发酵菌液100 mL,用氮气冲洗瓶内空间后,密封,隔绝空气,在45 ℃条件下发酵反应,每隔一段时间取气样分析,40 d完成全部试验。

## 2 实验结果

### 2.1 产气量

86 mL/g 糖蜜,总产气量约为 860 mL。主要气集中在反应的前 4 d,最高峰值约 300 mL/d。

### 2.2 气体组成分析

气体分析结果列于表 1 中。由于采样分析过程中混入少量空气,表 1 中的结果是去除空气后的数据。

**表 1 气体组成分析结果**

**Table 1 Analysis of composition of gases**

| No. of samples 取样编号 | Reaction time 反应时间(h) | Composition of hydrocarbon gases 烃类气体组成(%) | | | | | | | | $CO_2$ (%) |
|---|---|---|---|---|---|---|---|---|---|---|
| | | $CH_4$ | $C_2H_6$ | $C_3H_8$ | $iC_4H_{10}$ | $nC_4H_{10}$ | $iC_5H_{12}$ | $nC_5H_{12}$ | $C_6^+$ | |
| 1 | 6 | 98.09 | 1.39 | 0.07 | 0.17 | 0.08 | 0.14 | 0.05 | 未检出 | 29.21 |
| 2 | 12 | 92.93 | 3.66 | 0.40 | 0.22 | 0.99 | 0.52 | 0.87 | 0.40 | 58.59 |
| 3 | 29 | 92.47 | 3.19 | 0.50 | 0.42 | 1.63 | 0.74 | 1.05 | 未检出 | 79.70 |
| 4 | 36 | 96.50 | 0.40 | 未检出 | 未检出 | 1.40 | 0.69 | 1.01 | 未检出 | 78.43 |
| 5 | 95 | 92.54 | 0.13 | 0.38 | 1.79 | 1.38 | 2.07 | 0.37 | 1.31 | 68.65 |
| Average 均值 | | 94.51 | 1.75 | 0.27 | 0.52 | 1.10 | 0.83 | 0.67 | 0.34 | 62.92 |

注:$C_6^+$ 代表具 6 个碳原子和多于 6 个碳原子的烃类化合物之和。

## 3 讨论

### 3.1 烃类组成的特点

众所周知,石油烃类气体有自己独特的区别于其他成因、其他领域的气体的组成特点。从表 1 数据可见,微生物发酵产生的气体,从组成上来看,完全是真正石油伴生气的特点:

(1) 乙烷以上($C_2^+$)重烃类气体 >5%;

(2) 具有显著含量的 $C_6^+$ 烃类(平均 0.34%);

(3) $nC_4/iC_4>2$,符合沼气发酵液中脂肪酸含量及其变化关系;

(4) 从 $CH_4$ 至 $C_6^+$ 的烃类含量分布变化与石油伴生气完全一致,与沼气发酵液中脂肪酸含量变化关系及其脱羧规则吻合。

### 3.2 关于二氧化碳

大量资料表明,除个别油气田外,多数油溶气、石油伴生气中 $CO_2$ 含量均较低,而表 1 中的 $CO_2$ 含量却很高,似乎与“真正石油气”组成相矛盾。其实不然。

按笔者研究,“湿封闭体系”是饱含水的,且与碎屑沉积物及胶结物都有碳酸盐矿物存在,$CO_2$ 必定参与下列反应而减少甚至从气体产物中消失:

$$CO_2 + CaCO_3 + H_2O \rightleftharpoons Ca(HCO_3)_2$$

然而,$CO_2$ 含量偏高于一般沼气,原因尚待检查,但不影响本文的结论。

## 4 结论

从实验结果可见,由微生物发酵方法获得的烃类气体,具有显著量( >5% )的乙烷以上的重烃类气体,尤其是含有显著量(0. 34% )的石油中特有的 $C_6^+$ 重烃,加之丁烷异构体比值与岩石吸附烃相差甚远,而完全与石油伴生气一致,这均有力地表明石油烃类气体是由微生物直接将处于水体呈溶解状态的脂肪酸脱羧而成的。由此也不难看出,石油并非由细粒沉积物——过去称之为“油源岩”中的干酪根热降解形成的。

致谢:感谢大庆油田勘探开发研究院张春英高级工程师安排实验与分析;感谢浙江农业大学钱泽澍教授的指导与无锡市第二制药厂林锡能先生的帮助。

### 参考文献

[1] He Zhigao. On the origin of petroleum hydrocarbon gases[J]. Acta Geologica Sinica, 1986,60(4):57 -65.

[2] He Zhigao. A model of oil-generation in a waterlogged and closed system[J]. J Southeast Asian Earth Sci,1991,5(1 -4):465 -469.

[3] 钱泽澍,闵航. 沼气发酵微生物学[M]. 杭州:浙江科学技术出版社,1986.

# 13. 窄而厚的相变带沉积物富含石油——碳酸盐岩地区一种找油的重要领域

# Narrow and thick sediments of the "Phase Change Zone" is rich petroleum

何志高

(地质矿产部石油地质中心实验室,江苏无锡,214151)

碳酸盐岩地区的石油地质的特殊性制约着它的油气普查勘探方向。本文通过应用一种能反映碳酸盐岩特殊性的地球化学方法的研究,提出一种理论,即在碳酸盐岩地区,窄而厚的泥质碎屑沉积和碳酸盐沉积之间的相变过渡带具有生成巨量石油的条件,该过渡带可能是石油普查勘探的一个重要的领域。

## 1　方法

长期以来,传统的"有机碳—抽提物—干酪根"的研究方法对划分、评价、对比碳酸盐岩生油岩效果不佳,尤其是对中国中、古生代碳酸盐岩,因其演化程度高,保存条件差,受油气运移、富集的干扰,分不清岩石中烃类的原生性,故而其研究效果更差。

然而,业已发现,酸解气态烃对评价碳酸盐岩生油岩具有独特的意义,其方法具有显著的优点。[1]

酸解气态烃是指用酸分解碳酸盐矿物后所释放出来的烃类气体。通过对该类气体的地球化学特征的研究,认为它是一种不受油气运移、富集的影响,基本上不受演化程度影响的原生烃类。参考各类天然气(包括石油气、瓦斯气、沼气等)在组成上的特点,并依据地质剖面实测资料,提出了用酸解气态烃湿度($C_2^+ \times 100/\Sigma HC$)及含量作为生油岩的判别标准,并对比评价生油岩。该方法可以应用于地素样品、岩屑样品的分析研究。钻井以及在分析前

的全过程中无玷污干扰。其地球化学特征如下:

1.1 不受热演化程度的影响

所有钻井剖面样品分析的资料表明,酸解气态烃含量(μL/kg 岩石)及温度与井深无关,与样品热变质程度无关。

表1是5口井综合资料,由该表可见:上部石炭系样品的演化程度相对较低(干酪根 H/C 原子比较高,为0.86),酸解气态烃湿度较低,平均为1.55%;下部泥盆系的演化程度较深(H/C 原子比较低,为0.61),湿度较高,平均为4.8%,为上部石炭系平均值的3倍。由此可见,演化程度不影响湿度,不存在目前油气演化理论的烃类演化趋势。从表1还可以看到,同样均是灰岩,石炭系的 $c_1$、$d_1$ 样品,酸解气态烃总量只有2251 μL/kg,而泥盆系的邵6井 $D_2q$,气态烃总量为62580 μL/kg,相当于 $c_1$、$d_1$ 的27.8倍。可见,气态烃总量也不受演化程度的影响,不因演化程度高而损失烃的数量。

其他地区,如贵州羊昌河羊4井系统剖面以及“震旦-三叠系拼接大部面”的大量资料可见,酸解气态烃总量与湿度均不受演化程度的影响。

为了进一步证实该结论,我们还进行了模拟试验。试验结果发现,酸解气态烃在岩石中是相当稳定的,在400 ℃以下,其湿度基本不变,仅仅只有在500 ℃以上才有一定的变化。

**表1 湖南湘中坳陷5口井资料**

**Data of five well of Xiang-Zhong Depression Hunan**

| Well No | 地层 Strata | ΣHC (μL/kgrock) | 湿度 Humidity (%) | Kerogen | | 油气显示 Display of oil and gas |
|---|---|---|---|---|---|---|
| | | | | $\bar{R}_0$ | H/C | |
| S-1 | $c_1$ $d_1$ | 24 | 3.4 | >3 | 0.86 | |
| 3 | $c_1$ $d_2$ | 43873 | 0.7 | | | |
| 1 | $c_1$ $d_1$ | 2383 | 5.5 | | | 未见井下油显示,一口井有少量气,$CH_4$ 含量99%以上。地表有油苗。 |
| 3 | $c_1$ $d_1$ | 88807 | 1.0 | | | |
| 3 | $c_1$ $y_2$ | 86970 | 1.3 | 2.3 | 0.61 | |
| 1 | $c_1$ $y_1$ | 236 | 1.3 | | | |
| | Mean | | 1.55 | | | |

续表

| Well No | 地层 Strata | ΣHC (μL/kgrock) | 湿度 Humidity (%) | Kerogen | | 油气显示 Display of oil and gas |
|---|---|---|---|---|---|---|
| | | | | $\bar{R}_0$ | H/C | |
| S-8 | $D_3$ $X_3$ | 1199 | 7.4 | | | 未见井下任何油气显示。地表有油苗。 |
| | $D_3$ $X_3$ | 3125 | 1.1 | | | |
| | $D_3$ $X_2$ | 3411 | 1.1 | | | |
| | $D_3$ $X_1$ | 365 | 7.3 | | | |
| | $D_3$ S | 1607 | 10.2 | | | |
| S-4 | $D_2$ q | 1247 | 2.1 | >3 | 0.31 | |
| 6 | $D_2$ q | 62781 | 0.3 | >2 | 0.29 | |
| 6 | $D_2$ q | 558 | 6.0 | | | |
| | Mean | | 4.8 | | | |

### 1.2 不受油气运移、富集的影响

在碳酸盐岩样品中，常见到裂隙含油、晶洞含油或者油浸状含油灰岩以及很高的氯仿抽取物，这对评价样品的生油性能带来很大的困难。表2是一块湘中石炭系含油灰岩样品，油味很浓。从表2资料可见，岩石中含油系天然石油无疑。按照传统方法进行评价，当然这块岩石为生油岩。然而，酸解气态烃则含甲烷达99.9%，乙烷仅0.1%，含极少量的更重的烃类，是一种不同于石油气的干气的特征。可以认为，岩石中的石油是外部运移来的。这个结论可以从有机碳与氯仿沥青含量上相近的相互关系得到旁证。显然可以说，外部运移来的石油没有对酸解气态烃产生干扰。

此外，还进行了一些试验，如同样的样品，氯仿抽提前后进行酸解气态烃的分析、热解色谱分析等。结果表明，有机溶剂不渗透矿物晶体内部，从而影响气态烃的赋存状态，改变它的组成与含量。因此，酸解气态烃是一种特殊的赋存状态，不受油气运移和富集的影响，从而可以依据气态烃本身的组成与含量上的特征评价岩石的生油性能。

### 1.3 地表样与井下样可以对比

表3是同层位的地表、井下样分析资料的对比。由表3可见，区域上气态烃湿度最高的是 $T_2b^3$，其次是 $T_2b^1$ 与 $T_2f$，两类样品分析资料是一致的。说明本方法可适用于地表样品的研究，无疑这是相对于传统方法的一个优点。

表 2 湘中石炭系含油灰岩分析资料

| Corg (%) | $B_{CHF}$ (%) | 沥青族组分(%) | | | | 饱和烃色谱 | | |
|---|---|---|---|---|---|---|---|---|
| | | 饱和烃 | 芳烃 | 胶质 | 沥青质 | Pr/Ph | 主峰碳 | 碳数区 |
| 0.14 | 0.135 | 79.0 | 12.1 | 7.3 | 1.6 | 0.1 | 18 | 13～31 |
| 酸解气态烃分析 | | | | | | | 芳烃红外光谱 | |
| $CH_4$ | $C_2H_6$ | $C_3H_8$ | | $\Sigma HC$ (mL/kg) | | $C_2 \cdot 100/\Sigma CH$ | $A_{810}/A_{750}$ | |
| 99.9 | 0.1 | 痕 | | 109.1 | | 0.1 | 0.8 | |

表 3 地表样与井下样对比

| 样品 \ 数据 \ 项目 | | | 酸解气态烃(μL/kg) | | | | | | | |
|---|---|---|---|---|---|---|---|---|---|---|
| | | | $CH_4$ | $C_2H_6$ | $C_3H_8$ | $iC_4$ | $nC_4$ | $iC_5$ | $nC_5$ | 湿度 |
| $T_2b^1$ | 钻井 | 瘤状灰岩 | 6121 | 367 | 108 | 27 | 27 | 11 | | 8.1 |
| | 钻井 | 生物杂屑泥灰岩 | 16135 | 736 | 171 | 45 | 39 | 28 | | 5.8 |
| | 地表 | 紫红色瘤状泥灰岩 | 23192 | 3034 | 565 | 85 | 90 | 31 | | 14.1 |
| $T_2b^3$ | 钻井 | 含泥质灰岩 | 80067 | 26402 | 5837 | 1018 | 1036 | 428 | 223 | 30.1 |
| | 地表 | 泥灰岩 | 80656 | 35510 | 9699 | 1674 | 1378 | 590 | 187 | 37.8 |
| $T_2f$ | 钻井 | 浅灰色灰岩 | 7010 | 364 | 70 | 16 | 13 | | | 6.2 |
| | 地表 | 浅灰色生物灰岩 | 3890 | 359 | 73 | 12 | 9 | | | 10.3 |

### 1.4 气态烃各单组分含量变化的一致性

对湘中数口井的气态烃各组分含量变化进行相关性数理统计分析，其结果表明，同一口井的不同深度的样品，各单一组分含量之间具有良好的相关性，各单一烃类含量具有相似的变化趋势。

它们之间的线性相关系数值如表 4 所示。这种高的相关系数值，表明它们之间成因一致，后生变化无影响。按烃类的物理化学性质来说，它们之间的差异是极其巨大的。仅以沸点为例，甲烷与乙烷相差 73 ℃。若后生变化显著，其变化值不可能一致。由此可见，烃类的后生变化是不显著的。也就是说，后生变化(主要是热的作用的影响)对烃类气体的含量与组成无显著的影响，从而反映了酸解气态烃的原生性以及它的特殊的赋存状态。

**表4 气态烃各组成部分线性相关系数**

| $x$ \ $r$ \ $y$ | $CH_4$ | $C_2H_6$ | $C_3H_8$ | $iC_4H_{10}$ | $nC_4H_{10}$ |
|---|---|---|---|---|---|
| $CH_4$ | 1 | 0.9453 | 0.899 | 0.6578 | 0.6347 |
| $C_2H_6$ | | 1 | 0.9457 | 0.7991 | 0.6831 |
| $C_3H_8$ | | | 1 | 0.8506 | 0.8677 |
| $iC_4H_{10}$ | | | | 1 | 0.8517 |
| $nC_4H_{10}$ | | | | | 1 |

### 1.5 赋存状态

上述资料已经表明,酸解气态烃的湿度与样品的井深深度无关,不受演化程度影响,400 ℃以下不逸出,氯仿溶剂抽提不出来,都说明这类气态烃非岩石表面吸附,而是一种特殊的赋存状态。

通常的灰岩的孔隙度<1%,岩石的密度若以2.5计,每千克样品的孔隙体积约4 mL,而酸解气态烃的量可高达数十万微升,约为孔隙体积的百倍。可见气态烃绝不赋存于岩石孔隙之中,不是一种游离的、容易出出进进的、储集于孔隙或裂隙通道的气体,而是被牢牢地裹在碳酸盐矿物之中。只有将碳酸盐矿物分解后,它才裸露并释放出来。

## 2 生油岩标准

在综合我国中、古生代碳酸盐岩几条钻井剖面和部分地面样品以及松辽、泌阳等陆相碎屑岩样品的分析资料的基础上,参照各类天然气的成分特点和成因机制,提出利用酸解气态烃湿度($C_2 \cdot 100/\Sigma HC$)和气态烃含量作为划分和确定碳酸盐岩生油岩标准(表5)。

**表5 碳酸盐岩生油岩碳解气态烃标准**

| 类别 \ 指标 | | $\frac{C_2^+ \cdot 100}{\Sigma HC}$ (%) | $\Sigma HC$ (μL/kg) |
|---|---|---|---|
| 碳酸盐岩生油岩 | Ⅰ | >25 | |
| | Ⅱ | 25~15 | |
| | Ⅲ | 15~5 | >4000 |
| 非生油岩 | | <5 | <4000 |

湿度主要反映有机物类型,反映成油的倾向性。这方面前人已做过研

究,рогозина 从西伯利亚沉积岩样封闭孔隙中气体的组成研究得出结论:气体的成分直接与原始有机物质类型有关,即湿度愈高,有机物的类型愈有利于生油,详见表 6。

**表 6 气体组成与有机物成因类型对比 ***

| 有机物成因类型 | 气体组成(%) | | | | | 湿度(%) |
|---|---|---|---|---|---|---|
| | $CH_4$ | $C_2H_6$ | $C_3H_8$ | $C_4H_{10}$ | $C_5H_{12}$ | |
| 腐泥型 | 2.6 | 10.9 | 39.6 | 33.2 | 13.7 | 97.4 |
| 腐殖-腐泥型 | 54 | 14.6 | 27.1 | 3.7 | 0.6 | 46.0 |
| 腐泥-腐殖型 | 83.2 | 13.2 | 3.3 | 0.3 | 0.0 | 16.8 |

* adapted from рогозина

根据羊 4 井良好的油气显示以及气态烃湿度对比,划分了生油岩的类别。

表 5 中酸解气态烃的量主要是反映有机物丰度的,其界限值是初步确定值。根据研究,好的生油岩,其值是相当高的,愈高愈好,其上限尚难确定。但是一般来说,好的生油岩应该达 $5 \times 10^4 \mu L/kg$ 左右。

## 3 标准应用

将表 5 的标准应用于湘中、贵州、江苏、华北等地区碳酸盐岩,各类样品的达标率综合对比列于表 7 中。

**表 7 各地区样品达标率综合对比**

| 达标率(%) \ 标准 / 地区 | 有机碳(%) | | | 有机碳 0.10% 烃含量 60 ppm | 酸解气态烃 | 石油显示 |
|---|---|---|---|---|---|---|
| | 0.08 | 0.10 | 0.24 | | | |
| 湘中 274 个样品 | 94.5 | 89.4 | 40.5 | 0.73 | 1 | 很差 |
| 石炭系 | 100 | 94.4 | 27.8 | 0 | 0 | 很差 |
| 泥盆系 | 80 | 70 | 17.6 | 0 | 3 | 很差 |
| 贵州羊 4 井 | 100 | 100 | 64.5 | 94.7 | 100 | 很好 |
| 羊 5、7、8 井 | | | | | 82 | 好 |
| 拼接大剖面 | | | | | 27 | 较好 |
| 江苏(一口井) | | | | | 42 | 较好 |

由表 7 可见,对于油气显示很差的湘中地区来说,各种有机碳标准都是不

合适的。烃含量由于受运移与演化程度的影响，当然也是不合适的，而酸解气态烃标准达标率为1% ~3%，可以认为是在情理之中。

从贵州羊昌河（表8）的资料来看，羊4井湿度>25%的几组地层，是油气显示特别好的井段，如180 ~250 m、265 ~275 m、528 ~538 m等，标准与实际上的油气显示对应得如此之好，不能不认为是酸解气态烃标准有效性的良好检验。

拼接大剖面、江苏剖面都在一定程度上反映了新指标的有效性、合理性。

## 4　富油的岩相变化带

大量样品的分析资料指出，酸解气态烃的含量及湿度均与样品的岩性关系密切。

由表8可见，凡是油苗很多、油气显示好的井段，湿度高，岩性均是泥灰岩、泥质灰岩；而灰岩、泥岩井段，湿度低，油气显示较差。同样，拼接大剖面中样品的湿度与岩性的关系也是明显的。表9更直观地反映了岩性与气态烃含量之间的关系：黑色泥岩的碳酸盐矿物含量低，酸解气态烃总量很少，仅23.5 μL/kg；灰质泥岩的有机碳含量虽然达0.19%，但是气态烃总量仍十分低；而泥质灰岩（生-34），有机碳含量为0.22%，稍高于18号样品，但酸解气态烃总量却高达$1.55\times10^5$ μL/kg，相当于18号的660倍。

**表8　贵州羊昌河羊4井资料（Data of Young-4 well, Guizhou）**

| 井深 Depth (m) | 岩性 Lithic | ΣHC (μL/kg rock) | 湿度 Humidity | Corg (%) | 沥青B (%) | 烃 Hydro-carbon (%) | 油气显示 Display of oil and gas |
|---|---|---|---|---|---|---|---|
| 175 | 泥质灰岩 | 141506 | 30 | 0.38 | 0.042 | 0.027 | 187 ~205 m 油苗甚多 |
| 180 | 泥质灰岩 | 164804 | 32 | 0.32 | | | |
| 215 | 灰质泥岩 | 8199 | 21 | 0.42 | 0.047 | 0.026 | |
| 237 | 砾状灰岩 | 41228 | 17 | 0.37 | 0.0079 | 0.006 | |
| 265 | 泥质灰岩 | 104252 | 33 | 0.34 | | | 265 ~275 m 油苗特多 |
| 290 | 灰岩 | 54951 | 17 | 0.27 | 0.006 | 0.005 | |
| 308 | 泥质灰岩 | 144331 | 32 | 0.27 | | | |
| 321 | 泥灰岩 | 18348 | 21 | 0.20 | 0.047 | 0.033 | |

续表

| 井深 Depth (m) | 岩性 Lithic | ΣHC μL/kg (rock) | 湿度 Humidity | Corg (%) | 沥青 B (%) | 烃 Hydro-Carbon (%) | 油气显示 Display of oil and gas |
|---|---|---|---|---|---|---|---|
| 356 | 泥灰岩 | 12213 | 30 | 0.22 | 0.055 | 0.029 | |
| 366 | 泥灰岩 | 12081 | 20 | 0.28 | 0.032 | 0.019 | |
| 494 | 泥灰岩 | 54556 | 20 | 0.30 | 0.058 | 0.043 | |
| 510 | 灰质泥岩 | 23422 | 15.2 | 0.38 | 0.042 | 0.029 | |
| 562 | 泥质灰岩 | 77401 | 28.3 | 0.31 | | | 528~538 m 油苗特多 |
| 579 | 灰质泥岩 | 107683 | 28.4 | 0.43 | 0.024 | 0.021 | |
| 602 | 灰质泥岩 | 68194 | 19.0 | 0.35 | 0.047 | 0.037 | |
| 623 | 泥灰岩 | 40548 | 15.2 | 0.33 | | | |
| 640 | 泥质灰岩 | 80981 | 15.3 | 0.37 | 0.040 | 0.032 | |
| 661 | 泥质灰岩 | 26923 | 16.3 | 0.32 | | | |
| 668 | 泥质灰岩 | 41244 | 22.1 | 0.25 | 0.036 | 0.028 | |
| 715 | 灰质泥岩 | 75999 | 16.6 | 0.25 | 0.056 | 0.029 | |
| 749 | 泥灰岩 | 86395 | 21.7 | 0.37 | | | |
| 773 | 泥灰岩 | 80.989 | 21.7 | 0.36 | | | |
| 791 | 泥灰岩 | 91420 | 15 | 0.46 | 0.045 | 0.032 | |
| 842 | 泥灰岩 | 23464 | 21.1 | 0.36 | | | |
| 850 | 泥灰岩 | 54195 | 25 | 0.35 | 0.076 | 0.050 | 851~865 m 油苗甚多 |
| 881 | 灰质泥岩 | 30320 | 16.2 | 0.39 | 0.034 | 0.025 | |
| 902 | 泥质灰岩 | 33668 | 17.2 | 0.48 | | | |
| 924 | 泥质灰岩 | 59401 | 13.7 | 0.53 | 0.048 | 0.039 | |
| 951 | 灰质泥岩 | 8811 | 13.9 | 0.40 | | | |
| 966 | 泥灰岩 | 6450 | 11.9 | 0.36 | | | |
| 989 | 灰质泥岩 | 189754 | 27 | 0.43 | 0.011 | 0.006 | |
| 说明 | 1. 整个井油苗丰富,饱和烃色谱 OEP 平均 1.02,碳数分布范围 $C_{13}$ ~ $C_{30}$ 芳烃红外光谱 810/750　0.8~0.85。<br>2. 资料来源,湖北综合研究大队,贵州八普。 | | | | | | |

**表 9 气态烃含量与岩性对比**

| 样品号 | 井号 | 井深(m) | 岩性 | ΣHC (μL/kg) | CaO + MgO (%) | 有机碳 (%) |
|---|---|---|---|---|---|---|
| 补-22 | 邵 1 | 535 | 黑色泥岩 | 23.5 | 0.42 | 0.14 |
| 18 | 邵 1 | 446 | 灰质泥岩 | 235 | 0.84 | 0.19 |
| 34 | 邵 1 | 855 | 黑色泥质灰岩 | 2383 | 29.37 | 0.95 |
| 生-34 | 邵 3 | 493 | 泥质灰岩 | 155380 | 49.83 | 0.22 |
| 18 | 邵 3 | | 灰色灰岩 | 142 | 55.57 | 0.23 |
| 导-3 | 坑道 | 地表 | 黑灰色泥质灰岩 | 12399 | | 1.49 |

因此可以认为,在碳酸盐岩地区,纯的或含灰质很少的泥岩以及纯的碳酸盐岩的生油能力差,即使有较高的有机物质,生油能力并不强。表 1 所列的湘中石炭系碳酸盐岩,有机碳含量相当高,生油能力很差,就是属于生油能力差的典型。生油能力最强的是泥灰岩、泥质灰岩,表 8 所述的羊 4 井就是最佳的典型。

根据贵州八普资料,羊昌河位于中国贵州三叠系"S"形相变带的东段。在这个相变过渡带,泥质含量增多,分散有机质,生物碎屑丰富。羊 4 井即位于该过渡带的鼻状背斜的轴部。

羊 4 井从井深 175 m 至 989 m,累积厚度 800 m,从上至下,油苗丰富,油气显示好。在该井剖面上,每隔 25 m 左右,系统采样一块,按生油岩标准(表 5)判定,全部样品达标,达标率为 100%。再者,在剖面中,湿度高(即 $C_2^+$ 的重烃含量高)的样品对应地吻合于录井评价油气显示良好的层位,表明该井油气显示属原生性,是一种自生自储。

更重要的是,该相变带的油气储集量惊人地丰富。按体积法估算,其生油气量如表 10 所示,每单位(以平方千米计)生油量,非封闭性为 $6.4\times10^5$t。位于相变带上的构造,以长 20 km、宽 2 km 计算,非封闭性生油量为 $2.4\times10^7$t,封闭型为 $2\times10^7$t,总生油量为 $4.4\times10^7$t。

这里所划分的封闭与非封闭型是根据酸解气态烃方法原理将岩石中的石油分成不同赋存状态的两部分:非封闭型是指有机溶剂氯仿抽提出的可以形成油藏的石油;封闭型是指碳酸盐矿物被分解后释放出来的石油。

**表 10　羊昌河相变带生油(气)量的推算值**

| 范围 \ 类型 | 非封闭的 | | 封闭的 | | 总量 | |
|---|---|---|---|---|---|---|
| | 油(t) | 气($m^3$) | 油(t) | 气($m^3$) | 油(t) | 气($m^3$) |
| 单位($km^2$)量 | $6.4\times10^5$ | | $5\times10^5$ | $1\times10^8$ | $1.1\times10^5$ | |
| 相变带(20 km×2 km)量 | $2.4\times10^7$ | | $2\times10^7$ | $4\times10^9$ | $4.4\times10^7$ | |

## 5　结论与讨论

综上所述,通过采用新的地球化学方法研究,取得了对碳酸盐岩区生油有利带的认识,这个有利的生油带是相变过渡带。根据岩石物理性质以及岩石的结构分析可以看到,这种过渡带也是孔隙发育的良好渗透性的石油储集带,因此,不仅从生油,而且从储油的观点出发,相变过渡带对石油普查勘探是最有意义的,应该将石油普查勘探的重点放在这样一些富油的地区上。

Bailey Kascoe 等[2]研究报道了美国中大陆(Mid-Continent)的古生代碳酸盐岩油气田,其油田分布就在一条狭窄的从石灰岩向碎屑岩的过渡带上,其气田就在油田附近,其形状亦是条带状,只不过比油田要宽些。根据中国地质学家的研究(关士聪等),中国南方下古生代的大量油气苗、沥青矿亦是处于这样一类的过渡相带上。[3]

因此,本文从地球化学的观点与方法总结出来的规律性的认识,是有一定的地质实践资料验证的。

### 参考文献

[1] 何志高.应用酸解气态烃评价碳酸盐岩生油岩[J].石油实验地质,1983,4.

[2] Bailey Kascoe JR, Adler FJ. Permo-carboniferous hydrocarbon accumulations, Mid-Continet, USA[J].AAPG Bulletin, 1983,67(6).

[3] 关士聪,演怀玉,丘东洲,等.中国海陆变迁、海域沉积相及油气(晚元古代—三叠纪).北京:科学出版社,1984.

# 14. 试用油气成因新理论讨论台湾油气勘探选区与油气性质评价

# Try the new theory of the oil-generation to discuss exploration of oil-gas for the constituency in Taiwan

何志高，樊文玲

（地质矿产部石油地质中心实验室，江苏无锡，214151）

**提要**：湿的封闭体系综合成油理论揭示了石油形成的时间、地点、条件和方式，从成因上指出了油气聚集的最佳地质体，从而为台湾的油气勘探选区与原油性质评价提供新的理论依据。

基于此种构思，笔者提出值得注意的台湾油气勘探区及其目的层有：(1) 台湾西部上第三系上新统苗栗群海相沉积相变过渡带；(2) 台湾西部上第三系中新统煤层侧向变化领域；(3) 台湾中央山脉地层区中新统澳底组黑色页岩-煤的侧向变化区域。在油气性质评价方面，笔者认为，不能以干酪根Ⅲ型的成油母质类型归属油源及成油条件，而应根据原油中成分的地球化学特征来评价原油。

油气成因理论必定要指导、推动油气勘探的发展。本文根据油气成因新理论，分析研究台湾的石油地质资料及其勘探成果，提出某些油气勘探的有利地区与层位。同时，对油气性质进行油源归属及成油条件的研究。

因此，本文首先介绍油气成因新理论，然后在此基础上论述诸有利地区与层位的实际及原油性质，最后得出结论。

## 1 油气成因新理论

经过近20年的研究，笔者近来陆续发表了《论石油烃类气体成因》、《湿封闭体系综合效应成油论》以及《窄而厚的相变带沉积物富含石油一碳酸盐岩地区一种找油的重要领域》等论文，全面阐述了石油成因新理论。

该理论将成油机制分为两个阶段：第一阶段称为准备阶段，亦可称为形成甲烷与脂肪酸阶段；第二阶段称为成油-聚集阶段。

### 1.1 准备阶段

有机物（包括陆源的与水生的）随同各类碎屑物沉积后，在失去游离水前，受到生物化学作用及各种分异作用的影响，重新分配，最轻的、细分散状的以及胶体状的有机物聚集在细粒沉积物（如泥质）的表面，并和上部沉积覆盖的粗碎屑（如砂粒）相混，从而处于含水的多孔介质之中。这种赋存于孔隙中的有机物质在继续进行微生物发酵作用的时候，沉积覆盖另一个沉积旋回的细粒沉积物，此种沉积物作为盖层，从而形成一个隔氧的、饱含水的封闭体系。

在此饱含水的封闭体系中，有机物质经微生物的长期改造，形成各种脂肪酸类、类石油胶质、类石油沥青质以及甲烷等。此阶段类似于（但不同于）人工沼气发酵。这个阶段无生油作用发生，仅是一种生油前期的准备阶段。

### 1.2 成油-聚集阶段

在饱含水的封闭体系中，继形成脂肪酸的生化反应之后，开始了脂肪酸的脱羧生化反应。这种反应形成各种气态烃类及各种高分子烃类。这些烃类，加上胶质与沥青质，构成了早期生成的石油。这时期相当于碳酸盐岩的固结成岩期。

随着埋深作用增强及气态烃、可气化的烃等气体的生成，体系中的压力随之增大，温度增高。由于压力和温度的作用，分散的石油聚集成体积大小不等、含油级别不同的含油体。温度的另一种作用是，类石油胶质、类石油沥青质降解形成一部分烃类、胶质与沥青质。这部分烃类与脂肪酸脱羧形成的烃类构成了封闭体系中全部石油烃烃类之总和。

在高渗透性的饱含水的封闭体系中，所形成的气态烃、液态烃、胶质及沥青质在压力的作用下，气态烃进一步液化，石油各组分分子间内聚力增强，相互溶解性增强，黏度降低，运移能力增强，分散状的石油聚集成为一个整体，并受岩性等条件的影响，最终形成规模不等的油藏。

## 2 台湾油气有利地区（及地层）的预测

迄今，台湾陆区已发现铁砧山、畸顶、白沙屯、青草湖、锦水、竹东、六重溪、牛山以及高雄西面近海共 9 个气田；发现出磺坑、山子脚、宝山、竹头畸、冻子脚、台西共 6 个油田。油气田均集中在台西盆地，油气产层几乎全为中新统石英砂岩层。在海域已发现天然气、凝析油及原油，有的亦在开发之中。

笔者参考这些实际资料，并以此为基础，用新的油气成因理论提出自己的见解。

### 2.1 西部上第三系上新统苗栗群海相沉积相变过渡带

苗栗群，分布广泛，北起基隆，穿越西部平原直至高雄，是一套海相沉积，沉积厚度从北至南由 2000 m 增厚至 5000 m，粒度变细，至高雄有石灰礁出现。该地层分为下部的锦水组与上部的卓兰组。

该地层陆区未见产油报道，通常作为下伏地层的良好盖层，但南部台南坳陷砂岩为产气层（牛山、六重溪气田位于坳陷的北部）。

笔者认为，卓兰组为一套砂岩、粉砂岩与页岩互层，局部含灰岩礁，厚度可达 1500 ~ 2500 m 的沉积地层，其砂页岩互层本身是一套多组合的“封闭体系”。

按新理论，泥质碎屑中的有机质重新分配后进入上复砂层中，该砂层又有泥质覆盖，这种封闭体系对生油十分有利，而且砂层渗透性好，十分有利于石油的运移聚集；页岩中生物化石较多表明有机物质是比较丰富的，微生物对有机物质的改造作用是较强烈的，因此，通过重新分配作用进入砂体中的有机物质的数量比较多。

根据压力与温度效应，该地层组形成油藏的最大希望是埋藏深度比较大的地区。从深度意义上来说，在一定范围内，愈深愈有利于形成丰富石油的储油体。

依上述两原则，既要有富含有机物质的具有砂泥恰当比例（如 1∶1）的砂泥岩互层，又要有较深的埋藏深度。台湾西部山麓地区油气颇有希望的地区是卓兰组深度在 3000 m 以上的地区。

### 2.2 西部上第三系中新统南庄组及其煤层侧向变化领域

台湾西部上第三系中新统南庄组为一套含煤地层，煤层多。上部含煤 7 ~ 8层，其中 2 层可采；下部亦含煤数层，下层亦可采。

一般情况，煤层上下层位是富含有机物质的，煤层顶底砂岩见油的情况相当普遍。例如，四川香溪侏罗系煤层底部产油；浙江长兴-广德地区二叠系煤层顶底砂岩产油等。

广东茂名油页岩是众所周知的储量丰富的含油页岩，其油页岩从上向下取样岩石中有机碳含量逐渐增高，由 20% 增至 75% 以上，其底部完全是典型煤。而其含油率则由 15% 下降至 1% ~ 2% 。

某些有机地球化学家解释这种煤系地层产油的机制时，一则简单地说“煤可生油”，二则曰“煤系地层可以生油”，因为成煤物质中存在着诸如树脂类物质可以生油。

笔者认为,煤系地层在一定条件下确实可以生油。其条件是具有形成“封闭体系”的砂岩(或砂粒碎屑沉积物)或长期能“饱含水”的多孔隙,高渗透性的沉积物。其砂岩或高渗透性的沉积物可以造成微生物对有机物质的充分作用,形成脂肪酸类,进而形成烃类,再而,分散的微石油进行运移、聚集。

南庄组侧向变化的地层在岩性岩相上有粗粒沉积物,这种相变过渡带正是形成良好“封闭体系”的条件。因此,在这种相变过渡带进行油气勘查是颇有希望的。这种相变过渡带通常是比较窄的,如贵州的资料表明,该过渡带仅 1 ~ 3 km。

3. 台湾中央山脉地层区中新统澳底组黑色页岩-煤的侧向变化区域

澳底组建造于中新世早期,上部枋脚段为含煤段,其中有两层劣质煤。其层段厚度 350 m,该组总厚 720 ~ 1200 m(西亚区),含有孔虫及瓣鳃类化石。从生油条件上看,有机物质是比较丰富的,亦具备形成“湿封闭体系”的相变带,相变带的厚度较大,可能形成工业油气藏,因此,可以将澳底组的煤、黑色页岩的侧向岩相变化区域作为油气勘查的重点区域之一。

## 3 油气性质评价

中新统产出的油气性质特征已有资料介绍。笔者认为,在对油气性质的评价上,由于理论认识上的差异将导致油气远景评价、油源归属等重大问题的判别差异,因此,对这一问题的讨论无疑是有重要意义的。

虽然油气性质的分析资料并不丰富,但是油气的特征已基本上显示出来。这些特点可以归纳为以下几点:

(1) 白沙屯天然气烯烃含量较高, <2.2% 。

(2) 出磺坑原油芳烃含量较高,可达 55% (体积分数)。

(3) 锦水凝析油密度小,为 0.785 ~ 0.831;含硫量可达 0.05% ;含蜡量较低,为 0.029% ~10% ;芳烃含量较高,为 27% ~43% (体积分数);环烷烃含量较高,达 36% 。

原油一般为轻油,多汽油馏分和煤油馏分,在气相色谱图上,$C_{20}$以后的含量低。

我们首先讨论芳烃。按笔者对芳烃地球化学特征的研究,陆相原油芳烃含量很低。我国大庆油田,原油芳烃含量 < 10% ,泥岩的氯仿抽提物(沥青A)芳烃含量约 8% ;茂名油页岩抽提物约为 12% 。海相原油芳烃含量较高,一般可达 30% 左右。江苏南部二叠系海陆交互相地层的岩石氯仿抽提物芳烃含量出现较大的变化值域,介于 10% ~50% 。

很明显，芳烃的地球化学基本特征是陆相低、海相高。但是，无论陆相还是海相原油，轻质馏分芳烃较低，重质馏分中较高。如大庆原油，汽油馏分中芳烃<5%，煤油馏分稍高些，柴油馏分可达10%。

台湾锦水凝析油芳烃含量达30%～40%，属高芳烃含量，显然具有海相原油特征。因此，锦水凝析油不能简单地归属为Ⅲ型干酪根的产物。

其次，讨论硫含量。硫的基本地球化学特征是陆相（淡水）低、海相高，陆相膏盐沉积环境所形成的原油中硫含量很高，如潜江（江汉）原油，最高可达11%，也就是说，原油中硫含量的高低主要取决于沉积水体盐度。当然，原油的轻重不同馏分，硫含量亦存在着很大差异。对凝析油来说，几乎全是含硫量很低的。因为，硫多以高分子的非烃类化合物形式存在于油质中，在低分子的凝析油中必然不多。我国大陆陆相凝析油含硫量<0.01%，而锦水凝析油含硫0.05%，显然，锦水凝析油含硫量已经相当高了，不能认为是陆相原油特征，尤其不能认为是陆相淡水型原油特征，而正说明是海相环境的产物。

再者，从原油中环烷烃含量来看，环烷烃含量的高低是母质类型优劣的反映。笔者对环烷烃的地球化学研究结论是，随成油母质类型的变优，成油条件愈好，环烷烃含量愈高。如河南泌阳油田，通常认为具备最佳的生油条件，是“小而肥”的油田，其环烷烃含量可高达50%以上。锦水凝析油环烷烃含量高达36%，同芳烃的讨论结论一样，不能将油源归属为Ⅲ型干酪根。

天然气中烯烃的高含量特点可以认为是低成熟的反映。然而，烯烃几乎均系成气期间生物作用的产物，而非烃类气体和石油的热裂解产物。

## 4 结 论

根据上述讨论，台湾的原油油源不能归属为生油潜量甚低的Ⅲ型干酪根。锦水凝析油是一种生油条件甚佳、生油母质甚好的石油。

以新的石油成因理论为指导，在台湾可能会发现新的油气资源。尤其值得重视的是煤的侧向变化带以及泥质碎屑沉积与碳酸盐沉积之间的相变过渡带，但是这种相变过渡带通常是比较窄的。

### 参考文献

[1] 支家生，等. 台湾石油与天然气地质研究[M]. 杭州：〔出版社不详〕. 1987.

[2] 浙江石油地质研究所. 台湾石油地质通讯，1984(1)，1985(1)，1986(1).

[3] 何志高. 松辽及其他陆相盆地生油岩基本地球化学特征[J]. 石油实验地质，1981，3(3)：197－202.

[4] He Zhigao. On the origin of petroleum hydrocarbon gases. Acta Geologica Sinica, 1986,60(4):57-65.

[5] 何志高. 湿封闭体系综合效应成油论[C]. 全国第三届有机地化会议论文摘要汇编,1986.

本文原载于《台湾石油通讯》1988 年第 2 期

# 15. 台西盆地油气组成与性质差异的成因分析

## "Taihxi" basin oil and gas properties of the composition and nature of the cause of variance analysis

何志高

(地质矿产部石油地质中心实验室,江苏无锡,214151)

台湾台西盆地第三系油气田油气组成和性质的规律性变化及有关的石油地质问题早已有详细而深入的研究。根据天然气中 $CO_2$ 含量随产层深度的增加而增加的规律,普遍认为,台西盆地天然气中的 $CO_2$ 是由深处的石灰岩受岩浆热力作用所致,本区各组可能相互沟通,存在着油气(包括 $CO_2$)的垂向运移。笔者对此提出异议,认为产层中的 $CO_2$ 并非无机成因,而是有机成因;各上/下产层亦非相互沟通;油气无明显的垂向运移,有关油气组成、性质、生成和运移聚集等问题需要重新认识。

### 1　油气组成与性质变化规律简述

#### 1.1　青草湖气田

青草湖气田被称为台湾四大生产气田之一,面积约 4 $km^2$,发现于 1976 年。表 1(据支家生,1986)是该气田各产层油气性质的对比。从表 1 可见:

(1) 天然气,从浅层至深层,甲烷($CH_4$)与二氧化碳($CO_2$)含量增高,比重、发热率降低,而氮氧($N_2+O_2$)含量基本不变;

(2) 凝析油,从上往下,芳烃的含量、比重增大,硫含量亦略有增加;

(3) 天然气与凝析油的比重,从上往下的变化趋势恰好相反。

表 1　青草湖气田油气性质对比及垂向变化趋势

| 产层时代 | 产层名称 | 天然气 | | | | | | 凝析油 | | |
|---|---|---|---|---|---|---|---|---|---|---|
| | | 面深（m） | 比重 | $CH_4$（%） | $CO_2$（%） | $N_2+O_2$（%） | 发热率（kcal/$m^3$） | 比重 | 硫（%） | 芳烃（%） |
| 上新世 | 锦水段 | 1500 | 0.749 | 77.65 | 0.52 | 0.52 | 11514 | 0.712 | 0.02 | 6.0 |
| 上中新世 | 鱼藤坪段 | 1517 | 0.707 | 80.93 | 0.07 | 0.49 | 10980 | 0.724 | 0.01 | 10.03 |
| | 关刀山段 | 1800 | 0.697 | 83.19 | 0.95 | 0.52 | 10811 | 0.747 | 0.02 | 16.93 |
| | 东坑段 | 2243 | 0.710 | 80.14 | 1.57 | 1.55 | 10606 | 0.815 | 0.09 | 30.10 |
| 中中新世 | 打鹿段 | 2271 | 0.700 | 83.11 | 4.25 | 0.67 | 9992 | 0.787 | 0.04 | 23.0 |

### 1.2　台西盆地油气田

有资料表明，台西盆地除青草湖气田外的其他油气田油气性质与组成的变化关系同青草湖气田基本一致，其中二氧化碳含量的变化规律完全一致。

锦水气田：为陆区两大气田之一。从上中新统东坑段往深部至上渐新统五指山组，7 个段、组产出天然气的二氧化碳含量分别为：东坑段 1% ~2%，打鹿页岩段 3.26%，出磺坑组 3% ~9%，木山组 8% ~12%，五指山组 13% ~15%。

通宵-铁砧山气田：是陆区最大的气田。从中中新统打鹿页岩段往深部至五指山组，11 个段、组产出天然气的二氧化碳含量变化趋势与青草湖、锦水气田一致。打鹿页岩段 2% ~3%，往下逐渐增高，至五指山组达 37%。

出磺坑油气田：是台湾四大产气田之一。二氧化碳含量的变化规律与上述一样。以其中的木山组为例，第 6 产层（2892 m）为 36.6%，往下第 7 层至第 10 层依次为 39.30%、41.30%、43.30%、44.00%，从上往下增加，规律性极强。

畸顶气田、竹东气田：资料虽不多，但二氧化碳含量的变化规律在与横向上诸气田对应的产层对比时，仍可显示出与其他气田的一致性。

### 1.3　台南地区竹头畸油气田

该油气田在地理位置上与上述各油气田相距甚远，在构造上被前中新统基底隆起分隔。然而，该气田（产层为上中新统长枝坑组）天然气中二氧化碳含量（0.21%）与青草湖气田关刀山段地层（相邻接者）的含量（0.95%）完全可以对比。

## 2　油气组成性质差异的成因分析

在许多油气田中，天然气与原油性质之间、天然气中二氧化碳与烃类含量之间存在着密切的关系，这早已为许多资料所证明。但对其成因分析则各持己见。

### 2.1 大港油田和苏北油田的有关资料分析

大港油田、苏北油田均是第三系油田（指所研究的资料），它们的油气性质可供与台西盆地进行对比，如表 2、表 3 所示。表 4 是地层原油（高压物性取样器取样）油溶气组成随压力变化而受影响的试验资料。

**表 2 大港油田油气性质对比**（据《石油地质实验基础》，1978）

| 地层 | | 天然气 | | | 原油 | | | | |
|---|---|---|---|---|---|---|---|---|---|
| | | 比重 | 甲烷（%） | 重烃（%） | 比重 | 黏度（Pa·s，$\times10^{-3}$） | 凝固点（℃） | 胶质 + 沥青质（%） | 蜡（%） |
| 上第三系 | m-2 | 0.5734 | 96.7 | 2.1 | 0.9322 | 112.6 | -15.3 | 17.6 | 6.1 |
| | m-3 | 0.6093 | 93.5 | 6.6 | 0.9184 | 62.7 | -1.3 | 15.6 | 8.0 |
| | m-4 | 0.6626 | 85.0 | 12.9 | 0.8991 | 29.8 | 2.8 | 12.6 | 9.0 |
| | g-1 | 0.6568 | 85.6 | 12.0 | 0.8878 | 18.3 | 16 | 9.2 | 9.2 |
| | g-2 | 0.7081 | 77.0 | 17.4 | 0.8777 | 62.0 | 21 | 12.3 | 11.9 |
| | g-3 | 0.7316 | 76.0 | 22.0 | 0.8695 | 25.6 | 24.8 | 9.8 | 14.7 |
| | g-4 | 0.7827 | 71.4 | 27.7 | 0.8546 | 11.7 | 30.5 | 7.2 | 14.8 |
| 下第三系 | d | 0.7023 | 80.7 | 17.5 | 0.8411 | 4.99 | 25 | 8.67 | 16.37 |
| | S-2 | 0.6856 | 82.9 | 15.1 | 0.8583 | 7.59 | 21 | 9.66 | 18.26 |
| | S-3 | 0.7582 | 75.2 | 23.2 | 0.8280 | 2.87 | 23 | 5.66 | 12.49 |

从表 2 可见，从浅层至深层，天然气比重增加，重烃含量增高，甲烷降低；原油比重随深度增加而降低。这种油气性质之间的关系与青草湖气田的油气间的关系是完全相反的，即大港油田上部油重气轻下部油轻气重，而青草湖气田则是上部油轻气重下部油重气轻。这种对比可以表明，台西盆地各油气田在纵向上的天然气组成和性质的差异不是垂直运移所致，亦不能归属于热演化程度的差别，因为表 2 可作为油气热演化系列的参考标准，热演化的油气性质变化趋势与青草湖气田是相悖的。

从表 3 可见，苏北油田高压地层原油油溶气中二氧化碳含量变化幅度很小，没有随深度（温度和压力）增加而增加的变化规律，$CO_2/CH_4$ 比值变化亦不大；从表 4 可见，压力降低，分泌出来的油溶气中二氧化碳含量无显著变化趋势，$CO_2/CH_4$ 比值略有增加，表明二氧化碳在原油中的溶解能力比甲烷稍强或者只是一种过程平衡速度的差异。这就是说，在油气藏中，二氧化碳没有与烃类气体很大不同的独自的垂直运移性质。因此，台西盆地主油气层之下即使有大量的二氧化碳，也不会单独自下部油气产层向上运移，因此，二氧化碳含量随产层深度减小而降低的现象不会是二氧化碳的垂向运移所致。

**表3 苏北油田地层原油油溶气有关资料(He,1986)**

| 井号 | 取样井深(m) | 油层温度(℃) | 地层压力($\times10^5$ Pa) | 地层(符号) | $CO_2$(%) | $CH_4$(%) | $\frac{CO_2}{CH_4}$ | $C_2H_6$(%) | $C_3H_8$(%) | $iC_4$(%) | $nC_4$(%) |
|---|---|---|---|---|---|---|---|---|---|---|---|
| 真20 | 1950 | 68 | 193.1 | ES | 0.53 | 41.05 | 0.013 | 15.19 | 20.15 | 3.59 | 10.03 |
| 真20 | 2000 | 72 | 216.4 | $Ed^2$ | 0.81 | 65.14 | 0.012 | 7.84 | 11.40 | 2.12 | 6.36 |
| 真19 | 2200 | 78 | 215.3 | $Ed^2$ | 0.35 | 60.37 | 0.006 | 12.93 | 12.70 | 1.67 | 6.64 |
| 真49 | 2200 | 84 | 276.8 | $Ed^1$ | 0.25 | 69.15 | 0.004 | 12.85 | 9.11 | 1.19 | 4.16 |
| 真80 | 2800 | 100 | 344 | $Ed^1$ | 0.66 | 66.91 | 0.010 | 14.21 | 9.28 | 1.25 | 4.35 |

**表4 压力影响地层原油油溶气组成试验资料(He,1986)**

| 压力($\times10^5$ Pa) | $CO_2$(%) | $CH_4$(%) | $\frac{CO_2}{CH_4}$ | $C_2H_6$(%) | $C_3H_8$(%) | $iC_4$(%) | $nC_4$(%) | $\frac{C_2^+}{\Sigma HC}\times100$ |
|---|---|---|---|---|---|---|---|---|
| 120.8 | 0.97 | 78.74 | 0.012 | 9.50 | 5.02 | 0.5 | 1.31 | 17.2 |
| 100 | 0.98 | 78.23 | 0.012 | 10.11 | 5.17 | 0.5 | 1.49 | 18.2 |
| 76.7 | 0.91 | 75.71 | 0.012 | 12.02 | 6.34 | 0.6 | 1.85 | 21.6 |
| 52.0 | 0.93 | 71.70 | 0.013 | 14.05 | 7.90 | 0.79 | 2.14 | 25.8 |
| 36.8 | 0.96 | 64.39 | 0.015 | 17.98 | 10.60 | 1.04 | 2.95 | 33.6 |
| 24.0 | 0.99 | 52.58 | 0.019 | 22.44 | 15.28 | 1.45 | 4.66 | 45.5 |
| 14.0 | 1.03 | 36.73 | 0.028 | 26.40 | 22.62 | 2.52 | 6.62 | 61.3 |
| 7.2 | 0.66 | 2.32 | 0.28 | 16.96 | 38.00 | 7.05 | 20.97 | 97.3 |

此外，天然气中氧氮含量基本不变的情况表明，青草湖气田各气层本身的封闭性良好，没有地表水的侵入干扰，没有空气混入，没有微生物的分解氧化。

总之，从上述可见，台西盆地诸油气藏油气组成与性质的差异，不是垂向运移、油气热演化程度差异以及微生物的分解氧化的差异所致；各油气产层本身封闭性良好，油气无上下运移。

### 2.2 二氧化碳的剩余性及其含量的控制机制

根据湿封闭体系综合效应成油论（何志高，1991），一种形成烃类和二氧化碳的化学反应式为：

$$RCOOH \longrightarrow RH + CO_2 \quad (1)$$

即脂肪酸 RCOOH 脱羧基生成 $CO_2$ 和烃类化合物 RH。从这点出发，$CO_2$ 与

RH 是以 1 mol 对 1 mol 的关系，以同一反应而同时形成的“孪生子”。有油气生成就必有二氧化碳生成。

在湿封闭体系中，$CO_2$ 的来源除(1)式外，无疑还有成油准备阶段所形成和非脂肪酸脱羧成烃反应的其他生物化学反应所形成以及水体中溶解的 $CO_2$ 等。在良好的成油体系中，(1)式来源是主要的。

在沉积和成岩阶段，以胶结物形式或以碎屑形式存在的碳酸盐(矿物)是普遍的，加之有水的条件，成烃反应所形成的 $CO_2$ 必然会与碳酸盐(如碳酸钙)发生下列反应：

$$CaCO_3 + CO_2 + H_2O \rightleftharpoons Ca(HCO_3)_2 \tag{2}$$

另外，下列反应平衡式也必然存在而影响气相中 $CO_2$ 的含量：

$$CO_2 + H_2O \rightleftharpoons H_2CO_3 \tag{3}$$

由于(2)式向右方向进行，碳酸盐胶结物被溶解，所形成的碳酸氢钙以 $Ca^{2+}$、$HCO_3^-$ 形式转入水中，使有效孔隙度增大，岩层储集性能变佳。这样，油气生成反应为自己的储存创造了有利条件。另一方面，由于(2)式的进行，(1)式脂肪酸脱羧形成的 $CO_2$ 不断减少(甚至几近消失)，保证了成烃反应的进行(质量作用定律)。

然而，温度和压力均要对各反应产生影响，尤其是温度对(2)式反应的影响，压力对(3)式反应的影响。温度增高，$CO_2$ 绝对量和相对百分率均增加。

综合以上所述，3 个主要的反应式(1)、(2)、(3)式共同控制了 $CO_2$ 的绝对含量及它在天然气中的相对百分含量。这也就是 $CO_2$ 的剩余性及其含量的控制机制。

### 2.3 油气性质和组成差异成因机制的证据

#### 2.3.1 二氧化碳含量与砂岩储层物性的关系

通过对比天然气中 $CO_2$ 含量与砂岩储层物性，发现它们之间有着密切的关系，即 $CO_2$ 含量高，则砂岩物性差。例如，通宵-铁砧山气田和锦水气田的对比关系如表 5 所示。尤其是锦水气田出磺坑组 $CO_2$ 含量最低的第 16 产层有效孔隙度最高达 20%，关系极其明朗，即剩余的 $CO_2$ 愈多，(2) 式向右进行的程度就愈差，有效孔隙度也愈低。

**表 5　通宵-铁砧山和锦水气田 $CO_2$ 含量与产层物性对比表**(据支家生，1986)

| 气田 | 产层 | $CO_2$ 含量(%) | 有效孔隙度(%) |
|---|---|---|---|
| 通宵-铁砧山 | 打鹿组 | 2～3 | 16～19 |
| | 木山组 | 5～15 | 8～11 |
| | 五指山组 | 37 | 6 |

续表

| 气田 | 产层 | $CO_2$ 含量(%) | 有效孔隙度(%) |
|---|---|---|---|
| 锦水 | 东坑段 | 1~2 | 5~26 |
| | 出磺坑组 | 3~9 | 6~20 |
| | 木山组 | 8~12 | 5~8 |
| | 五指山组 | 13~15 | 2~4 |

2.3.2 储层砂岩中碳酸盐胶结物含量与孔隙度的关系

根据储油岩资料,有效孔隙度与钙质含量的变化方向相反。实际上,这种关系可以说是砂岩储层的普遍规律。

2.3.3 同层位(或同时代)二氧化碳含量的可比性

台西盆地各油气田在横向上相同产层的 $CO_2$ 含量具有相当强的可比性,甚至可以将 $CO_2$ 作为“气体分子化石”来对比产气层的地层。这一特性是任何来源于产层下部无机成因的 $CO_2$ 向上垂直运移的观点难以解释的。

2.3.4 油与气二者产量的相互关系

台西盆地或其中单个油气田的油气产量统计表明,油与气二者在产量上的关系是油多气多,油少气亦少。在年度产量变化曲线上,油与气两条变化曲线十分相似,油的曲线上升,气的曲线也上升;油的曲线下降,气的曲线也下降。大家知道,在统计产量时,气的产量中是包含着 $CO_2$ 的。如果 $CO_2$ 是从下部基底或无机成因而来,就不会出现“气多油亦多、气少油亦少“这种油气紧密相依的关系。这从另一侧面说明了气中的 $CO_2$ 不是来自深部的无机成因源,也不存在油气的垂向(从下部产层向上部产层)运移,同时也说明了烃类与 $CO_2$ 的同生性。

## 3 结论

(1) 台西盆地各油气田天然气中 $CO_2$ 的相对百分含量随深度增加而增加的规律不能简单地解释为 $CO_2$ 的绝对量增加;不能因为纵向的这种从下往上、由高变低的分布特征说明油气存在垂直向上的运移和上下产层间相互沟通。

(2) 台西盆地各区气田天然气中所含的 $CO_2$ 是有机成因,它具有与原油及烃类气的同生性和与碳酸盐反应后的剩余性。

(收稿日期:1992-03-29)

## 参考文献

[1] 支家生.台湾石油天然气地质[J].台湾石油地质通讯,1986(4):1-43.

[2] 支家生.台湾石油地质勘探概述[J].台湾石油地质通讯,1984(1):3-18.

[3] 鞠天吟,陈其爽.台湾地层[J].台湾石油地质通讯,1985(2):4-25.

[4] 周次雄,等.台湾西部所产天然气中高含量二氧化碳来源之初步研究[J].台湾石油通讯,1991(15):23-26.

[5] He Zhigao. On the origin of petroleum hydrocarbon gases[J]. Acta Geologica Sinica, 1986,60(4):57-65.

[6] He Zhigao. A model of oil generation in a waterlogged and closed system[J]. J Southeast Asian Sci,1991,5(1-4):465-469.

本文原载于《台湾石油通讯》1992 年第 2 期

# 16. 天然气组成特点及其成因机制分析

# Composition characteristics of natural gas and the mechanism of its genesis

何志高

(地质矿产部石油地质中心实验室,江苏无锡,214151)

地质体中的天然气组成复杂多变,导致其成因研究困难。然而,长期积累的大量分析资料已为理论上科学论证提供了可靠的事实依据。笔者以热力学理论为基础,联系天然气的实际组成,研究可能生成烃类天然气的四大类型以及三种亚类型的化学反应,并系统地对各类型反应进行热力学分析。同时,在研究方法上是求反应产物(烃类气体)的相对丰度比值,而不是求单一烃类化合物的转化率及反应产物的分压等参数,从而近似地消除了地质体中极为复杂的平衡体系中的温度与压力等因素的影响。另外,在求比值的数学方法上是取地质体中的极限条件,消元解方程,近似求解。因此,复杂的化学问题以及地球化学问题得到了极大程度的简化,进而能够从理论上阐明多种天然气的成因。在掌握了组成与成因的规律后,后生变化影响因素的研究就会容易得多,这是十分有利于油气勘探与开发的。再者,热力学理论在石油地质中的应用也得到了进一步的发展。

## 1 天然气组成特点

### 1.1 油田气

油田气(包括与石油有成因关系的所有石油天然气)最显著的特点是组成上的高湿度($C_2^+ \cdot 100/\Sigma HC$),即乙烷以上的重气态烃含量比较高。

地面原油中呈溶解状态的气体中甲烷含量很低,湿度可高达99%以上,例如德克萨斯原油[1],甲烷含量仅0.01%。苏北油田地层原油放空后,即压力与大气压力相等时立即取样分析,甲烷为7.9%,湿度为92.1%。地层原油的气体烃类的湿度与压力的关系极大,随压力增大而湿度降低,如苏北油田周22井原油试验，压力从120.8个大气压降至7.2个大气压，其湿度由21.3%增至97.7%。油田顶部气的湿度可以很高，也可以较低，如大庆油田约为20%，苏北黄4井为7%。凝析油田气，湿度与油田顶部气近似，如马洛沙[2]油田，深度7000 m，压力1500个大气压，气体湿度为25.3%（摩尔百分数），四川凝析油气体湿度介于5%～20%。

在低压油溶气中还将出现 $C_2 < C_3$ 的情况。

### 1.2 煤田气

煤田气的组成特点是低中等湿度。根据现有资料，除个别特例外，大部分煤田气湿度小于5%。如内蒙古伊深1井石盒子煤样[3]，湿度为4.9%，江苏宜兴煤以及通常的煤田瓦斯气中甲烷含量高达99%，湿度相当低。湿度的变化主要与煤质有关。

### 1.3 沼气与浅层气

沼气的湿度几乎毫无例外地小于0.1%[3]。根据我们的研究，湿度小于0.01%。

沼气是生物成因气，但与通常的浅地层生化气是有区别的，成分上不可直接对比。从实际分析资料来看，通常的地层生化气湿度亦小于1%。例如，大部分的上海第四纪天然气与长江三角洲天然气湿度小于1%。但也有一部分天然气湿度是较高的，个别可高达8%。

### 1.4 高温与深层气

众所周知，泥火山气体中，乙烷以上重气态烃的含量很低。深部地层产热解气，湿度很低。例如，湖南石炭系泥灰岩产气层的重烃含量极低，仅0.002%（地矿部中南石油地质局）。

## 2 天然气成因的热力学对比分析

天然气的生成实际上是一种化学反应。众所周知，不同类型的反应物，不同的反应条件，均可产生不同的反应产物。这是研究天然气成因与组成关系的基础。

在化学领域中，热力学对反应的分析研究是严格的。地质体中的高温高压以及由此而发生的相态变化等，使得这种本来就要求严格科学分析的反应更加复杂化，研究的难度也更高。本文试图用一种简捷的方法，即化学-地球化学相结合，妥善处理变量之间的数学关系，仅求烃类化合物的相对丰度比值，从而使复杂的问题得到了极大程度的简化。

下面分析有关的四种主要类型的化学反应。

2.1　第一类反应

$$C_2H_4(g)+2H_2(g)\rightleftharpoons 2CH_4(g) \quad (1)$$

$$(1)式\ \Delta G^0=-40.562\ \text{kcal}\cdot\text{mol}^{-1}$$

$$C_2H_4(g)+H_2(g)\rightleftharpoons C_2H_6(g) \quad (2)$$

$$(2)式\ \Delta G^0=-24.14\ \text{kcal}\cdot\text{mol}^{-1}$$

$$根据\qquad -\Delta G^0=RT\ln K_P \quad (3)$$

按理想气体以及标准状态处理，将可进行下列数学演算：

令 $K_{P11}$ 为(1)式反应的平衡常数，$K_{P12}$ 为(2)式反应的平衡常数，$P_{Ma}$、$P_{Ea}$、$P_{Ee}$、$P_H$ 分别代表甲烷气、乙烷气、乙烯气、氢气的分压。则有

$$K_{P11}=\frac{P_{Ma}^2}{P_{Ee}\cdot P_H^2} \quad (4)$$

$$K_{P12}=\frac{P_{Ea}}{P_{Ee}\cdot P_H} \quad (5)$$

$$P_H=\sqrt{\frac{P_{Ma}^2}{P_{Ee}\cdot K_{P11}}}=\sqrt{\frac{P_{Ea}}{P_{Ea}\cdot K_{P12}}} \quad (6)$$

$$\frac{P_{Ma}}{P_{Ea}}=\frac{P_{Ee}^{\frac{1}{2}}\cdot K_{P11}^{\frac{1}{2}}}{P_{Ee}\cdot K_{p12}}=\frac{P_{Ee}^{\frac{1}{2}}}{P_{Ee}}\cdot K=\frac{K}{P_{Ee}^{\frac{1}{2}}} \quad (7)$$

据(1)、(3)式，得　　$\lg K_{P11}=29.745$

则　　$K_{P11}=5.56\times10^{29}$

据(2)、(3)式：　　$\lg K_{P12}=17.702$

则　　$K_{P12}=5.038\times10^{17}$

$$K=\frac{K_{P11}^{\frac{1}{2}}}{K_{P12}}=1.48\times10^{-3}$$

因此，

$$\frac{P_{Ma}}{P_{Ea}}=\frac{1}{P_{Ee}^{\frac{1}{2}}}\times1.48\times10^{-3} \quad (8)$$

取极限条件：

当 $P_{Ee}=1$ 个大气压时，$\frac{P_{Ma}}{P_{Ea}}=1.48\times10^{-3}$

当 $P_{Ee}=1000$ 个大气压时，$\frac{P_{Ma}}{P_{Ea}}=4.7\times10^{-5}$

当 $P_{Ee}=10^{-8}$ 个大气压时，$\frac{P_{Ma}}{P_{Ea}}=14.8$

这里应该说明，反应物与反应产物的分压取极限条件值时，由于 $\Delta G^0$ 的数

据是气体分压各自为1个大气压时所进行反应的自由焓变化值，所以不能以小于或大于1个大气压的条件计算平衡常数。否则，必须对温度、压力进行校正。然而，正如前言中指出的那样，我们采用的是一种求相对比值的方法，主要判别发展变化的趋势，而且，该值近似校正后，对结论不产生影响。因此，我们仍直接采用简捷的方法予以讨论。

在烃类的热裂解反应中，乙烯是主要的产物之一。以地层压力1000个大气压为极限，并根据波义耳定律进行计算，$CH_4/C_2H_6$ 仅 $4.7\times10^{-5}$，这种气体组成特征肯定不属石油类气体的特点。当 $P_{Ee}$ 等于1个大气压时，$CH_4/C_2H_6$ 比值为 $1.48\times10^{-3}$，虽然这种组成的气体在采出的地面原油中可以见到，但必须指出，地面原油中甲烷含量很低的情况是由甲烷的溶解度等性质所决定的，并非成因所致。当 $P_{Ee}$ 为 $10^{-8}$ 时，$P_{CH_4}/P_{C_2H_6}$ 比值接近石油气的成分，然而，如此低压稀少的反应物是不可能形成当今有开采价值的石油天然气藏的。鉴此，可认为第一类反应不是形成石油类天然气的反应类型。同时，由此分析也就表明了石油类气体不是由烃类裂解反应所形成的。当然，生物作用也能形成乙烯，若有 $H_2$ 伴生，形成少量的乙烷也不能排除。不过，生物成因的乙烯量通常是很少的，不足以形成大量的石油类天然气。

### 2.2　第二类反应

$$CO_2(g)+4H_2(g)\rightleftharpoons CH_4(g)+2H_2O(l) \tag{9}$$

(9)式 $\Delta G^0=-27.16\ \mathrm{kcal/mol}$

$$2CO_2(g)+7H_2(g)\rightleftharpoons C_2H_6(g)+4H_2O(l) \tag{10}$$

(10) 式 $\Delta G^0=-37.9\ \mathrm{kcal/mol}$

按第一类反应处理，并令 $K_{P21}$ 为(9)式平衡常数，$K_{P22}$ 为(10)式平衡常数，$P_C$ 为二氧化碳气体分压，则有：

$$K_{P21}=\frac{P_{Ma}}{P_C\cdot P_H^4},P_{Ma}=P_C\cdot P_H^4\cdot K_{P21}$$

$$K_{P22}=\frac{P_{Ea}}{P_C^2\cdot P_H^7},P_{Ea}=P_C^2\cdot P_H^7\cdot K_{P22}$$

$$\frac{P_{Ma}}{P_{Ea}}=\frac{K_{P21}}{K_{P22}}\cdot\frac{1}{P_C}\cdot\frac{1}{P_H^3}=K_2\cdot\frac{1}{P_C}\cdot\frac{1}{P_H^3} \tag{11}$$

根据计算　$$K_2=1.48\times10^{-8} \tag{12}$$

通常认为该类型反应是沼气形成的主要反应之一。根据模拟实验，通常 $P_C/P_H$ 比值可确定中值为0.5，则综合(11)、(12)式可得：

$$\frac{P_{Ma}}{P_{Ea}}=2.8\times10^{-8}\times\frac{1}{P_H^4} \tag{13}$$

人工沼气生产的实践指出，乙烷含量很低，均 $<0.1\%$。各取 $P_{Ma}/P_{Ea}=1000$，则有：

$$P_H=7.3\times10^{-2} \tag{14}$$

理论计算与实践相符，其氢气的浓度(分压)是一致的。式(11)指出，氢气浓度愈高，乙烷的生成量越多，气体愈湿。关于氢气浓度的这种规律性的结论，对以后讨论石油成因是颇有裨益的，此处不作深入讨论。

2.3 第三类反应

2.3.1 第一亚类反应

$$2C_6H_6(g)\rightleftharpoons 9C(\text{石墨})+3CH_4(g) \tag{15}$$

(15)式 $\Delta G^0=-67.41\ \text{kcal/mol}$ (16)

$$C_6H_6(g)\rightleftharpoons 4C(\text{石墨})+C_2H_6$$

(16)式 $\Delta G^0=-38.849\ \text{kcal/mol}$

同理，按第二类反应处理，则：

$$\frac{P_{Ma}^3}{P_{Ea}}=0.86\times10^{21}$$

当 $P_{Ea}=1$ 个大气压时，则 $\dfrac{P_{Ma}}{P_{Ea}}=0.95\times10^7$

当 $P_{Ea}=1000$ 个大气压时，则 $\dfrac{P_{Ma}}{P_{Ea}}=0.95\times10^5$

当 $P_{Ea}=0.001$ 个大气压时，则 $\dfrac{P_{Ma}}{P_{Ea}}=0.95\times10^8$

由此可见，该亚类型反应，$C_2H_6$ 分压越小，比值越大。$C_2H_6$ 分压在 0.001 ~ 1000 个大气压范围变动，其 $P_{Ma}/P_{Ea}$比值介于 $10^5\sim10^9$。

因此，地质体中无论甲烷与乙烷的分压各是多少，该类型反应的甲烷与乙烷比值($CH_4/C_2H_6$)均大于 $10^5$。

2.3.2 第二亚类反应

$$C_{17}H_{36}(l)\rightleftharpoons 8C(\text{石墨})+9CH_4(g) \tag{17}$$

(17)式 $\Delta G^0=-122.45\ \text{kcal/mol}$

$$C_{17}H_{36}(l)\rightleftharpoons 5C(\text{石墨})+6C_2H_6(g) \tag{18}$$

(18)式 $\Delta G^0=-60.35\ \text{kcal/mol}$

同理，按上述方法处理，则有：

$$P_{Ma}^{\frac{3}{2}}/P_{Ea}=3.86\times10^7$$

若 $P_{Ea}=1$ 个大气压，则 $P_{Ma}/P_{Ea}=1.14\times10^5$

若 $P_{Ea}=1000$ 个大气压,则 $P_{Ma}/P_{Ea}=1.14\times10^4$

若 $P_{Ea}=0.001$ 个大气压,则 $P_{Ma}/P_{Ea}=1.14\times10^6$

所以,该亚类反应可以认为 $CH_4/C_2H_6>10^4$。

2.3.3 第三亚类反应

$$C_{11}H_{24}(l)\rightleftharpoons 5C(\text{石墨})+6CH_4(g) \qquad (19)$$

(19)式 $\Delta G^0=-78.27$ kcal/mol

$$C_{11}H_{24}(l)\rightleftharpoons 3C(\text{石墨})+4C_2H_6(g) \qquad (20)$$

(20)式 $\Delta G^0=-36.87$ kcal/mol

按上述方法处理,则有:

$$P_{Ma}^{\frac{3}{2}}/P_{Ea}=3.86\times10^7$$

当 $P_{Ea}=1$ 个大气压时,则 $P_{Ma}/P_{Ea}=1.14\times10^5$

$P_{Ea}=1000$ 个大气压时,则 $P_{Ma}/P_{Ea}=1.14\times10^4$

$P_{Ea}=0.001$ 个大气压时,则 $P_{Ma}/P_{Ea}=1.14\times10^6$

所以,该亚类反应 $P_{Ma}/P_{Ea}>10^4$。

2.4 第四类反应

$$C(\text{石墨})+2H_2(g)\rightleftharpoons CH_4(g) \qquad (21)$$

(21)式 $\Delta G^0=-12.14$ kcal/mol

$$2C(\text{石墨})+3H_2(g)\rightleftharpoons C_2H_6(g) \qquad (22)$$

(22)式 $\Delta G^0=-7.86$ kcal/mol

同理,按上述类型反应处理,则有:

$$P_{Ma}^{\frac{3}{2}}/P_{Ea}=1.08\times10^{12}$$

当:$P_{Ea}=1$ 个大气压,则 $P_{Ma}/P_{Ea}=1.05\times10^6$

$P_{Ea}=1000$ 个大气压,则 $P_{Ma}/P_{Ea}=10^7$

$P_{Ea}=0.001$ 个大气压,则 $P_{Ma}/P_{Ea}=1.05\times10^9$

因此,可以认为该类反应 $P_{Ma}/P_{Ea}>10^7$。

## 3 综合分析与结论

由上述热力学与气体组成的分析表明,第一类型反应显然不是生成石油类气体的反应。第三类型中的一、二、三亚类反应是目前石油演化流行的反应模式,这种"演化论"认为,石油(包括长链饱和烃)受热分解,首先变成湿气,然后进一步变为甲烷干气。根据我们的试验和分析,石油烃类化合物,无论是长链烷烃、短链烷烃还是芳烃，热解反应（除模拟试验的特殊条件外）的产物绝大部分是甲烷，其 $CH_4/C_2H_6$ 比值可达到 $10^4$ 以上，并非湿气的组成。因此，也可以说，石油类湿气从成因上来说不是由石油烃类热解而成的。第四类反应，碳与氢的直接反应，从上述分析可见，其湿度更低，当然也就不会有形成石油烃类气的这种成因反应类型。也不难得出结论，石油烃类气体不是由碳与氢直接化合而成的。

第二类型反应，从上述反应可见，其产物组成变化的大小取决于二氧化碳与氢气的浓度。同时，二氧化碳与氢气二者之中，更主要的起决定性作用的是氢气浓度（或分压）。从反应式可见，$P_{Ma}/P_{Ea}$比值是变量氢气分压四次方的倒数的函数。氢气浓度扩大 10 倍，$C_2H_6$ 相对 $CH_4$ 的量扩大 10000 陪，$P_{Ma}/P_{Ea}$比值相应变小 1/10000。

从现有资料来看，氢气的地球化学研究尚不充分，难于作出肯定的结论。应该看到，甲酸的脱羧反应可以提供氢的来源。然而，从发酵有机物产生的沼气组成来看，$C_2H_6$ 含量仅为 $10^{-6}\sim10^{-5}$（体积分数），与第二类反应的理论计算值一致。因此，石油类天然气与沼气组成差别甚大，不能归属为目前的生物发酵反应形成。

（收稿日期：1984 年 12 月 26 日）

**参考文献**

[1] Welte D. Organic geochemistry of cardon[M]// Wedepohl KW, ed. Handbook of Geochemistry, 1969: 11/16L.

[2] Zeikus JG. The biology of methanogenic bacteria[J]. Bacteriol Rev, 1977, 41(2): 514 – 541.

[3] Larsen JW, Samst TL, Rodgers BR. Internal rearrangement of hydrogen during heating of coals with phenol[J]. Fuel, 1981, 60(4): 335 – 341.

[4] 何志高. 应用酸解气态烃评价碳酸盐岩生油岩[J]. 石油实验地质, 1983, 5(4).

[5] 魏永成，等. 四川六地区沼气组分的考察[M]//沼气（第四集）. 重庆：科学技术文献出版社重庆分社, 1980.

# Composition characteristics of natural gas and the mechamism of its genesis

He Zhigao

(The Central Laboratory of Petroleum Geology, Ministry of Geology and Mineral Resources)

**Abstract**: Based on the thermodynamic theory and combined with the composition characteristics of natural gas, four types and three sub-types of chemical reactions which might form hydrocarbon gases are studied. Thermodynamic analyses of various kinds of reactions are presented systematically. Meanwhile chemical and geochemical methods are used together not only to study the transformation ratio or the condition of the reactions, but also the relative abundance ratio of hydrocarbon compounds. The mathematic relationships among variables are properly treated. Thus, it simplified the complicated chemical and geochemical problems in geological body.

本文原载于《石油实验地质》1985 年第 4 期

# 17. 从热力学理论讨论烃类的演化

## Discussion on the evolution of hydrocarbons in the light of the theory of thermodynamics

何志高

(地质矿产部石油地质中心实验室,江苏无锡,214151)

地质体中的石油处于一种特殊的介质内,在漫长的地质历史时期,由于埋藏深度的增加,受着较高的温度和压力的影响,发生着改变石油成分和性质的种种演化。可以说,压力与温度是影响石油性质的主要因素。

石油成分的变化是一种化学变化。而热力学在解决化学变化方向上的正确性,已为许多化学实践所证实。本文以热力学理论为基础,侧重于压力因素的定性对比,结合我国湘中地区古生界碳酸盐岩等方面的资料,探讨石油的演化。初步研究表明,地层压力是石油演化中的一个重要因素。

### 1 压力是烃类热解反应的阻力

根据大量的化学、化工实验资料,烃类的热解产物主要为低分子烯烃与低分子烷烃。考虑到地质体中烃类的分布以及目前可能取得的热力学参数,确定热解反应的反应物以正丁烷为主要代表,生成物的低分子烃类为甲烷。在此基础上,分别讨论焓、熵、自由焓等热力学函数与压力之间的关系,进而讨论能量平衡对比,阐述地层压力是烃类热解反应的阻力。

#### 1.1 压力与焓

焓随压力改变而改变。通常的化学研究领域大多是比较低的压力,其影响较小,故而对压力不太重视。石油地质体中的地层压力是相当高的,以Malossa(意大利)油田[2]为例,油层压力高达1000个大气压以上。我国陆相中新生代油田的压力也一般都在100个大气压以上。

表1是烃类在500 K时不同压力条件下的焓值[1],以 $\Delta H_f$ 表示。$\Delta \overline{H}_p$ 称作平均变化率,以下式表示:

$$\Delta \overline{H}_p = \frac{\Delta H_{fT}^n - \Delta H_{fT}^1}{\Delta H_{fT}^1} \times 100 \quad (1)$$

式中：$\Delta H_{fT}^n$——某一压力下的焓值；$\Delta H_{fT}^1$——1 个大气压下的焓值。

**表 1　不同压力下 500 K 时烃类的焓值表（kcal · mol$^{-1}$）**

| 烃 | 类别 | 压力（大气压） | | | | |
|---|---|---|---|---|---|---|
| | | 1 | 10 | 100 | 200 | 300 |
| $CH_4$ | $\Delta H_{fT}$ | -18.6306 | -18.6451 | -18.781 | -18.902 | -18.981 |
| | $\Delta \overline{H}_p$ | | | -0.8 | | -1.9 |
| $nC_4H_{10}$ | $\Delta H_{fT}$ | -33.528 | -33.659 | -34.722 | -36.556 | -36.639 |
| | $\Delta \overline{H}_p$ | | | -3.5 | | -9.3 |

（此表根据参考文献[1]资料编制）

由表 1 可见，随压力增高，单个烃类化合物的焓值降低。降低的幅度，不同烃类是不一样的，正丁烷降低的幅度比甲烷大。用热力学理论来解释，焓值降低愈多（绝对值愈大），表明该化合物形成时放热更多，即这个化合物处于相对稳定的状态。由此看来，在一定压力下，从较高烃类产生甲烷的反应是不易进行的，因为压力在该反应中使反应物更趋稳定，从而阻止了反应的进行。

### 1.2　压力与熵

玻尔兹曼以下式定义熵：

$$S = K \ln Q \quad (2)$$

式中：$K$——玻尔兹曼常数；Q——分子运动混乱程度。

由于分子运动的混乱程度随压力的增加而降低，故从（2）式可知，压力增加，熵值降低。表 2 则是 500 K 时烃类在不同压力下的熵值。表中的相对变化率可用下式表示：

$$相对变化率 = \frac{S_p - S_1}{S_1} \times 100\%$$

式中：$S_p$——某一压力下烃类的熵值；$S_1$——1 个大气压下烃类的熵值。

表 2　500 K 时不同压力下烃类的熵值表(cal · deg$^{-1}$ · mol$^{-1}$)

| 烃 | 类别 | 压力(大气压) | | | | |
|---|---|---|---|---|---|---|
| | | 1 | 10 | 100 | 200 | 300 |
| $CH_4$ | $S$ | 49.479 | 44.879 | 40.088 | 38.505 | 37.531 |
| | 相对变化率(%) | | | -18.98 | | -24.15 |
| $nC_4H_{10}$ | $S$ | 89.074 | 83.833* | 76.05 | 74.16 | 73.31 |
| | 相对变化率(%) | | | -14.62 | | -17.69 |

* 12 个大气压下的熵值　　(此表根据参考文献[1]资料编制)

由表 2 可见,烃类熵值随着压力的增加而降低。但降低的幅度因烃类的不同而有差异,甲烷大于正丁烷。根据乙烷、丙烷的资料,压力对不同碳数烃类的熵值均有一定程度的影响,而相对变化率最大的是甲烷。因此,不难看出,烃类热解生成甲烷时,压力对该反应的产物(甲烷)影响最大,即生成物的熵值降得最低。根据熵增原理,降低熵值显然不利于该反应的进行。同理,在一定压力下,由较高碳数的烃类产生甲烷的反应也是不易进行的。

### 1.3　压力与自由焓

众所周知,在特定条件下,化学反应是朝着自由焓(能)降低的方向进行的。当体系的温度为零开氏度时,自由焓与焓相等。随着温度的升高,熵随之出现,并逐渐增大,自由焓与焓之差值亦随之出现并增大。因此,自由焓 $G$ 被定义为:

$$G = H - TS \tag{3}$$

式中:$H$——焓; $T$——绝对温度;$S$——熵。

根据前述压力与焓、压力与熵的讨论,通过(3)式可判定压力对自由焓的影响及影响方向。由自由焓与焓的差值可看出,压力是烃类热解反应的阻力。

表 3 是不同压力下 500 K 时烃类的生成自由焓 $\Delta G_f$。表中的相对变化率可用下式表示:

$$相对变化率 = \frac{(\Delta G_f)_p - (\Delta G_f)_1}{(\Delta G_f)_1}$$

式中: $(\Delta G_f)_p$——对应压力下烃类的生成自由焓;$(\Delta G_f)_1$——1 个大气压下烃类的生成自由焓。

**表 3　不同压力下 500 K 时烃类的生成自由焓表（kcal · $mol^{-1}$）**

| 烃 | 类别 | 压力（大气压） | | | | |
|---|---|---|---|---|---|---|
| | | 1 | 10 | 100 | 200 | 300 |
| $CH_4$ | $\Delta G_f$ | −7.8391 | −5.552 | −3.264 | −2.564 | −2.141 |
| | 相对变化率% | | 29.0 | 58.4 | 67.3 | 72.7 |
| $nC_4H_{10}$ | $\Delta G_f$ | 14.5348 | 16.946 | 18.553 | 18.964 | 19.305 |
| | 相对变化率（%） | | 16.6 | 27.6 | 30.5 | 32.8 |

（此表根据参考文献[1]资料编制）

表 3 的数据与热力学第二定律基本方程式（式 3）的结论是一致的。压力增加，自由焓增高。在不同烃类的变化率中，甲烷最大。这就表明，在烃类热解反应中，压力导致生成物的自由焓增大的幅度大于反应物自由焓增大的幅度，由此可得出压力是该反应的阻力的结论。

### 1.4　热解反应方向与能量的综合解析

烃类热解反应的产物随热解条件的变化而有很大差异，甚至同一条件下也难于得到相同组分的反应产物。况且，地层中所进行的反应与实验室模拟，化工产品的生产之间有着巨大的差异。本文以当前石油演化学说的热解反应为依据，研究气态正丁烷分解为气态丙烯和气态甲烷的反应。

表 4 是正丁烷在各种不同压力、温度条件下分解反应的能量分布与转化温度。高压条件下的 $\Delta G$ 值是能量平衡条件下的计算值，只能在表 4 范围内对比。转化温度 $T_{tu}$ 是指反应自发进行的最低温度的极限温度，是 $\Delta G$ 小于或等于零时的温度。

**表 4　不同压力、温度下正丁烷分解反应的能量分布及转化温度表**

| 反应条件 | | $\Delta H$ | $\Delta S$ | $\Delta G$ | $T_{tu}$（K） |
|---|---|---|---|---|---|
| 压力（大气压） | 温度（K） | （kcal/mol） | （cal · $mol^{-1}$ · $K^{-1}$） | （kcal/mol） | |
| 1 | 400 | 17.16 | 34.23 | 3.45 | ≈501 |
| 1 | 500 | 17.02 | 33.88 | 0.08 | ≈502 |
| 100 | 500 | 18.00 | 27.25 | 4.375 | ≈660 |
| 300 | 500 | 19.74 | 23.72 | 7.884 | ≈832 |

（此表根据参考文献[1]资料编制）

按照热力学理论，自由焓是化学平衡移动的动力，其微分式是：

$$dS-\frac{dQ}{T}\geqslant 0 \text{ 或}$$

$$TdS-dQ\geqslant 0 \tag{4}$$

在恒温恒压下,(3)、(4)两式综合可得(5)式:

$$-\Delta G=T\Delta S-\Delta H\geqslant 0 \tag{5}$$

表4中有一个较大的 $\Delta G$ 值,要使 $\Delta G\leqslant 0$,必须使式(5)中的 $T\Delta S$ 项等于或小于 $\Delta H$,从这点温度 $T$ 开始就出现反应的自发进行,故称此温度为转化温度。

由表4可见,随压力增高,正丁烷分解反应的焓增大,熵减小、自由焓增大,转化温度增高。根据汤姆逊-柏特洛原则,分解反应的焓增大,反应不易进行。根据熵增原理,熵减小亦不利于反应的进行。自由焓增大,从热力学基本方程式讨论,更不利于反应的进行。而能量变化的结果必然是转化温度增高。由此可见,压力是烃类热解反应的阻力。

## 2 热力学理论在石油地质中的应用

上述讨论表明:烃类热解形成甲烷气的反应,随着地层压力增高,从能量上来说,受到一种阻力。这种阻力如同反应中的能量壁垒,若要把反应向生成甲烷气的方向推进,使反应自发进行,必须提高温度。而通常地层压力所产生的阻力是相当大的,要达到克服阻力,逾越其能量壁垒,所需的温度是相当高的。例如,当压力达到100个大气压时,转化温度约为660 K(380 ℃);压力为300个大气压时,转化温度约为830 K(550 ℃)。这种高温对石油地质体来说是难于达到的,然而表4中的压力条件却是常见的。因此,在地质体中,诸如正丁烷之类的较低分子的烃类要热解成甲烷气的可能性,对广泛分布的石油来说是很小的。这也就是在烃类演化程度较深的地区还有原油存在的原因。现以我国湘中地区及意大利马洛沙凝析油田的资料来验证上述热力学理论的结论。

根据湖南石油会战指挥部资料,湘中地区中泥盆统碳酸盐岩的烃类地球化学特征表明,本区烃类的演化程度已相当深。但在道县螺海,却还有以“桶”度量的原油存在。该原油的族组分分析列于表5中。从原油芳烃红外光谱(图1)、非烃红外光谱(图2)、芳烃核磁共振波谱(图3)以及色谱分析(图4)可以看出:在演化程度较深的地区,不但有低分子烷烃,也有高于 $C_{20}$ 以上的正烷烃,还有支链相当长的多环芳烃以及高环环烷烃,而且还有富含杂原子的非烃与沥青质。

**表 5　湘中螺海原油族组分分析表**

| 样品＼组分 | 高于 300 ℃馏分 | | | | 低于 300 ℃馏分(%) |
|---|---|---|---|---|---|
| | 饱和烃(%) | 芳烃(%) | 非烃(%) | 沥青质(%) | |
| 原油 | 53.14 | 5.00 | 0.94 | 0.18 | 40.0 |

（石油地质中心实验室分析）

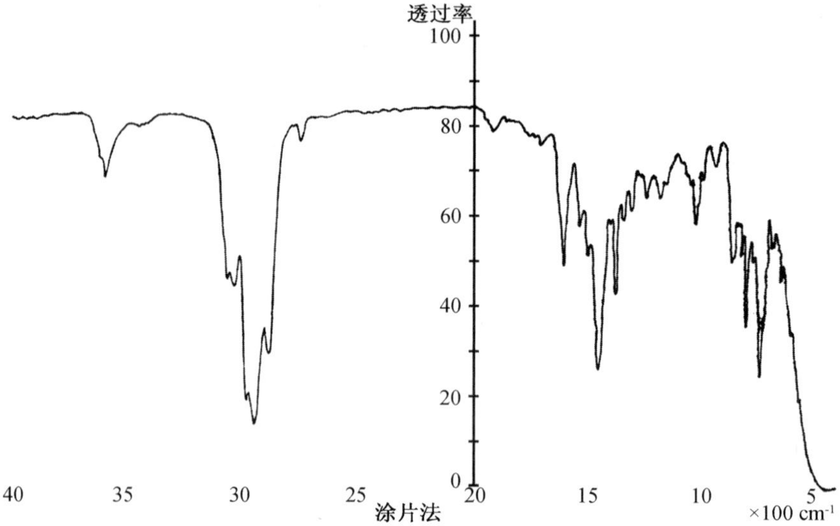


**图 1　道县螺海原油芳烃红外光谱图**

有人认为，油苗是因局部地区相对低温而被保存下来的。然而，湘中地区大量钻井岩样的分析表明，同一块样品，干酪根分析是高演化程度，而可溶抽提物中还有一定数量的饱和烃，这种饱和烃的色谱图，基本形态与螺海原油(图 4)相似，完全是一种石油类饱和烃的特征。因此，烃类的存在是广泛的，并非限于局部。

除湘中地区外，意大利的马洛沙油田也是一个油气演化的典型。该油气藏是欧洲最深的烃类的生产地，是世界上正式开发的、储层具超高压的三个油田之一。在极端的高压(15000 磅/吋$^2$)高温(155 ℃)条件下，保存着巨量的石油烃类。在该油田的石油中，不但有轻烃类存在，也有一定量的重烃类($C_{13}$)，还有少量的沥青质存在。

加勃里埃梁[3]研究指出，石油演化达到一定的阶段后，它的性质保持稳定不变，即与深度和温度的继续增加无关。研究成果汇集于表 6 中。根据他的文章，在 1700 ~ 5440 m 范围内，镜煤反射率大于 7.7，个别样品达9.7，最高煤变质阶段划在焦煤以上(相当于我国煤牌号瘦-贫煤阶段)，这个范围就是他所圈定的石油性质稳定不变的阶段。

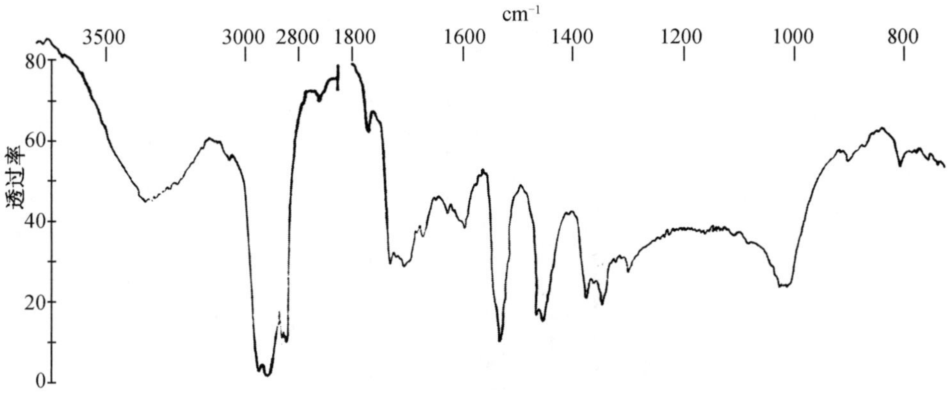


图 2　道县螺海原油非烃红外光谱图

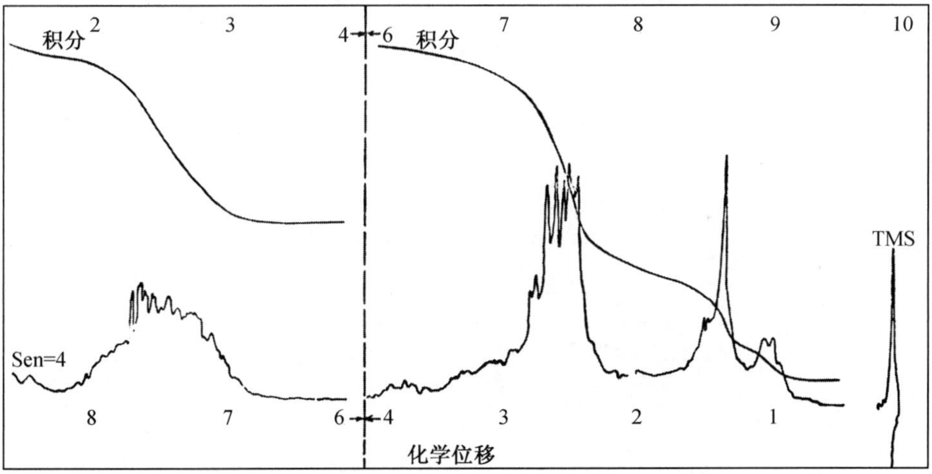


图 3　道县螺海原油芳烃核磁共振波谱图

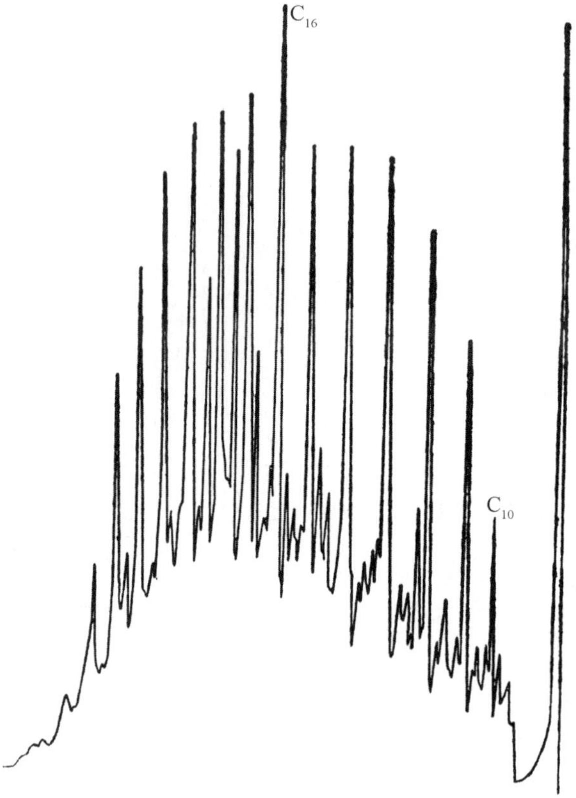


图 4　螺海原油色谱图(无机盐柱 FID)

**表 6　原油性质与深度的关系**

| 类别＼油田与时代 | 科罗勃科夫下石炭统 | 南乌麦特下石炭统 | 库季诺夫上泥盆统 | 乌斯提-波戈日上泥盆统 | 阿恩季波夫上泥盆统 | 乌斯提-波戈日中泥盆统 |
|---|---|---|---|---|---|---|
| 深度(m) | 1760 | 2650 | 3000 | 3820 | 4760 | 5200 |
| 密度($g/cm^3$) | 0.817 | 0.820 | 0.822 | 0.822 | 0.804 | 0.828 |
| 石蜡(%) | 2.76 | 1.40 | 4.20 | 7.00 | 5.70 | 7.5 |
| $S$(%) | 0.29 | 0.34 | 0.20 | — | 0.15 | — |
| $N$(%) | 0.11 | — | 0.04 | — | 0.04 | 0.05 |
| 小于 200 ℃ | | | | | | |
| 芳烃(%) | 7.0 | 8.0 | 9.0 | 6.0 | 8.0 | 15.0 |
| 环烷烃(%) | 39.0 | 43.5 | 34.0 | 47.0 | 43.0 | 33.0 |
| 烷烃(%) | 54.0 | 48.5 | 57.0 | 47.0 | 49.0 | 52.0 |

通过对上述地质体中有关烃类的地球化学的讨论认为，在演化程度很深的地区广泛存在着原油、岩石中存在高分子烃类的原因是，由于地层压力的增加，烃类不可能全部分解为甲烷气，从而经历很长的地质年代而被保存下来，形成油藏。

热力学的理论在它的产生、发展的历史中，无数化学上的事实证明是正确的。笔者通过对石油地质体中烃类热解反应的理论和实践的探索认为，热力学理论可以更深入地应用到石油地质中。

## 3 几点认识

### 3.1 烃类的汽化与演化假说

上述结论指出，低分子烃类热解成甲烷气的规律在石油地质体中不具普遍性。然而，多年来的油气普查勘探实践反映两个事实：一是地层埋藏愈深，时代愈老，石油愈轻；二是石油的储量、产量或者油田的数量，愈深愈老者愈少，而气则反之。可以说，深处油少气多。这种事实，过去人们认为，由于温度升高或者时间延长，石油被分解为甲烷气了。笔者认为，这是一种错觉，本文对此提出新的理论解释。

地质体中的石油，当演化到非烃被分解、沥青质被沉淀、成分变轻时（如马洛沙凝析油），烃类主要是汽化，而不是分解。烃类将按它们各自的饱和蒸气压力的大小，达成相态平衡。

阿伏伽德罗定律指出，在标准状态下，1 mol 气体体积约为 22.4 L。若以正戊烷为例，1 mol 液态其体积约为 110 mL，汽化成气态后，体积约为 22400 mL，体积膨胀了约 200 倍。若该容器（或构造圈闭）体积不变，在$PV = 1$（不讨论临界体积），体系的压力将增加 200 倍。

对一个油藏来说，烃类汽化，体积剧增，从而大大地提高了油藏圈闭系统的压力。油藏内压力的增加，导致封闭更加困难，微小的构造变动对油藏都是一种威胁。由此不难理解，油藏越深，温度越高，汽化的油越多，内压增加也越多，从而油藏被破坏的概率越大。因此，随着深度的增加，油藏则愈来愈少。

气藏则不同，温度改变时，气体压力改变较小，没有油藏那种相变存在，只有热膨胀。而这种热膨胀一般来说温度每增加一摄氏度，只改变其原来体积的1/273，对应的压力也只改变原来的 1/273。可见，这种改变比之于油藏因相变所造成的体积和压力的改变要小得多。由此可见，油藏受温度、压力的影响较大，易于遭到破坏；而气藏则受压力、温度的影响较小，不易遭受破坏。

### 3.2 油气普查勘探方向

上述烃类的热力学性质以及控制油气藏分布的规律,对于油气普查勘探具有重要意义。笔者根据这些理论对油气普查勘探提出以下几点意见:

(1) 高演化程度地区找到油的可能性虽较小,但只要保存条件好,例如有好的构造圈闭存在,依然可找到油藏,并非油都被热解成甲烷气。

(2) 一个地区的天然气藏(这里指的是干气)必将有它自己的独特成因,在勘探部署上不要与油直接联系而混为一谈。

(3) 大的构造圈闭虽然容易遭受破坏,但也具有有利于石油保存的一面。因为圈闭愈大,内压增加就少,因此,大圈闭有可能保存油藏。

热力学研究固然涉及很多领域,然而在石油地球化学的研究中却只是初步探索。由于有机物在地质体中的变化极其复杂,烃类热力学性质的参数不易取得,加之相态变化时能量的改变也很复杂,不易阐述清楚。本文仅以类比法,定性地通过逻辑推理来进行探讨,从比较中求认识,在此抛砖引玉,与同行们共同探索。

本文承孙肇才等同志审阅并修改,谨表谢意。

(收稿日期:1980 年 12 月)

**参考文献**

[1] Selected values of properties of hydrocarbons and related compounds. American peteroleum institute research project 44 volume Ⅰ、Ⅱ、Ⅲ、Ⅳ、Ⅴ.

[2] Chiericl GL, et al. Exploration and production in the very deep Malossa field[C], 10th World Petroleum Congress, 1979.

[3] 加勃里埃梁. 伏尔加河下游地区的区域变质及储集岩和石油的特征[J]. 石油地质科技情报,1976(2):40 –46.

[4] Greenwood HJ. 热力学在岩石学和矿床学的应用[M]. 曹万木,陈昆林,译. 北京:地质出版社,1976:40 –46.

[5] 邵美成. 熵及其应用[J]. 化学通报,1974(4).

# Discussion on the evolution of hydrocarbons in the light of the theory of thermobynamics

He Zhigao
(The Central Laboratory of Petroleum Geology, Ministry of Geology and Mineral Resources)

**Abstract**: Basing on the theory of thermodynamics, the influence of formation pressure on the parameters of hydrocarbons, such as enthalpy, entropy and free enthalpy, and the influence of pressure on the energy distribution in accord with the basic formula of thermodynamics $\Delta H = \Delta G + T\Delta S$ are discussed. It is pointed out that pressure is a negative factor to the pyrolysis of hydrocarbons. The reason for why there is oil still retained in Paleozoic carbonate rocks of the Xiang-Zhong district is also discussed under thermodynamic theory.

According to the theoretical consideration of the gasification of oil, as low boiling-point fractions in an oil pool increasingly gasified, the probability of the destruction of the oil pool also increased. Contrarily, there is no phase change within gas pools, only the expansion of gas under heating, hence the influence of temperature and pressure on gas pool is minimized. Therefore, it is easier for gas pools than for oil pools to be preserved. Finally, some cognitions for the prospecting and exploration of areas bearing highly evolved oil and gas are suggested.

本文原载于《石油实验地质》1982 年第 1 期

# 18. 关于沉积地球化学相的指示标志——铁的还原系数

# About sedimentary geochemical indicators—Iron reduction coefficient

何志高

（第三普查勘探大队实验室）

目前，在沉积地球化学相和生油层的研究中，应用着为数甚多的指示标志。其中对沉积环境和成岩过程物理化学特征的主要因素氧化-还原状况的研究比较重视。在表明氧化-还原状况方面，又以对铁的自生矿物组合系数（一般均称为铁的还原系数）的研究最为重视，尤其在郑爱珠等通过理论上的探讨和实践证明，改进了 A. И. 齐诺维也夫的 $K_{Fe}$ 系数，从而提出 $K$ 值系数以后，更加引起了国内相当多的石油地质工笔者的巨大兴趣。最近，又有研究者提出对 $K$ 值系数中的"0.236"改为"0.25"的意见，称之为 $K_1$ 系数。

笔者认为，在铁的还原系数问题上很值得探讨，对解决石油生成条件和指导油气勘探的理论和实践都有着重要的意义。

在铁的还原系数 $K_{Fe}$、$K$、$K_1$ 等系数中，有的是完全错误的，有的则存在着严重错误。本着求知和推动我国石油地质事业向前发展的愿望，抱着实事求是的态度，一方面批判这些错误；另一方面提出较为合理的系数来。

## 1 关于铁的还原系数 $K_1$ 值

按 $K_1$ 值（$K_1=\dfrac{Fe^{2+}_{HCl}\times 0.25+Fe^{2+}_{FeS_2}}{\Sigma Fe}$）的创笔者说：从理论意义上和实际显示规律性上来说，$K_1$ 值完全和 $K$ 值（$K=\dfrac{Fe^{2+}_{HCl}\times 0.236+Fe^{2+}_{FeS_2}}{\Sigma Fe}$）相似，但较 $K$ 值更为合理和易于推导，其理由主要有下述两点：

（1）将菱铁矿（$FeCO_3$）缩小至1/4，即可相当于黄铁矿（$FeS_2$）的耗碳量，这是一个当量折算值。因为按化学反应式来看,1 mol 的有机碳($C_{OP\Gamma}$)可以还原1 mol 的$FeCO_3$,同样,1 mol 的$C_{OP\Gamma}$却可还原4 mol 的黄铁矿铁。也就是说,生成菱铁矿的耗碳量大了4倍。

(2) 0.25 的来源也可如此计算:

$$0.25=\frac{0.053\cdots\cdots \text{生成 1 g } FeS_2 \text{ 之铁所消耗的 } C_{CP\Gamma}}{0.215\cdots\cdots \text{生成 1 g } FeCO_3 \text{ 所消耗的 } C_{CP\Gamma}}$$

所以,应将$0.236\times Fe^{2+}_{HCl}$改为$0.25\times Fe^{2+}_{HCl}$。

笔者认为,在这两条理由的第一条理由中,既然认为生成$FeCO_3$的耗碳量大4倍于生成$FeS_2$的耗碳量,那么,为什么不将$FeCO_3$乘以4,或者将$FeS_2$乘以1/4,而反去缩小菱铁矿之值,岂不是自相矛盾吗?从理论上来讲,更与事实不符,错误明显,怎么可能生成$FeCO_3$的耗碳量大于生成$FeS_2$的耗碳量。

在第二条理由中,若将计算式与"0.236"的计算式相对照,从原则意义上来说,恰好是分子分母的倒置:

$$0.236=\frac{0.215\cdots\cdots \text{生成 1 g } FeCO_3 \text{ 所消耗的有机碳的克数}}{0.91\cdots\cdots \text{生成 1 g } FeS_2 \text{ 所消耗的有机碳的克数}}$$

$$0.25=\frac{0.053\cdots\cdots \text{生成 1 g } FeS_2 \text{ 铁所消耗的有机碳的克数}}{0.215\cdots\cdots \text{生成 1 g } FeCO_3 \text{ 所消耗的有机碳的克数}}$$

如果按照这条理由,从意义上来分析,"0.25"需乘以$Fe^{2+}_{FeS_2}$,而不是乘以$Fe^{2+}_{HCl}$。这样的做法和上述一样,也是自相矛盾的,也是错误的。

当然,在提出$K_1$值的时候,还说了一些"$FeCO_3$是指铁元素,$FeS_2$是指全部分子式"的理由。不过,这种理由改变不了这一原则性的错误。况且,"0.236"的本身正反映了$FeCO_3$的"全部分子式",也反映了$FeS_2$的"全部分子式";恰恰相反,"0.25"才反映了$FeCO_3$的"全部分子式",而未将$FeS_2$的"全部分子式"反映出来。

"0.25"和"0.236"相差不大,当然可能在实际应用中显示出基本规律相似的情况。但是,在意义上却有着原则上的分歧,$K_1$值的理论依据更是有本末倒置之错。无可置疑,不能将"0.236"改为"0.25"。

## 3 关于 *K* 值中的"0.236"

郑爱珠等改进了齐诺维也夫的$K_{Fe}$值($K_{Fe}=\frac{Fe^{2+}_{HCl}+Fe^{2+}_{FeS_2}}{\Sigma Fe}$),在铁的还原系数中,分别对待菱铁矿和黄铁矿的观点是正确的,这两种形态的铁,在应用中的确不应该等量齐观。

但是,$K$值存在着相当严重的错误。本文主要侧重于从论述$FeS_2$的生成

及其中硫的性质和 $FeCO_3$ 的生成等问题入手，以物理化学的基本理论作为指导，来批判齐诺维也夫引用的几个化学反应方程式和“0.236”。

2.1 黄铁矿($FeS_2$)的形成及其硫的性质和“0.91”

齐诺维也夫撰写了一篇为表明岩石中有机碳和各种形式铁之间关系的文章，其中引用了几个化学反应方程式，这些反应方程式也广泛见于国内外有关石油地质学、地球化学等文献中。齐诺维也夫所提出的 $C_{Fe}^{S}=0.91$，就是依据这些反应方程式推导出来的；“0.236”也就是这些反应方程式的继续发挥。

齐诺维也夫按照 $Fe_2O_3$ 向 $FeS_2$ 的转变，从而计算出 $C_{Fe}^{S}$ 值：

(1) $2Fe_2O_3+C=4FeO+CO_2$， …………………………………… ①

$$\frac{C}{4Fe}=0.053;$$

(2) $SO_4^{2-}+2C=S^{2-}+2CO_2$， …………………………………… ②

$$\frac{2C}{S}=0.75$$

$$C_{Fe}^{S}=0.053+0.75\times1.14$$

$$=0.91 \qquad \text{③}$$

笔者认为，从这两个方程式本身来看，并无异议，应该承认它们反应进行的可能性。特别是第一个方程式(即方程式①)，从理论上来推断，在客观实际的变化中是更为确切的。但是，问题就在于将这两个方程式组合起来作为形成 $FeS_2$ 的定量计算的依据就是非常错误的了。其所以错误，中心问题在于将 $S^{2-}$ 与正 $[S_2]^{2-}$ 混淆起来，两者都作为 $H_2S$ 形态的硫来对待的观点和做法的错误。

综合各种资料、文献，推测 $FeS_2$ 的形成有以下几种可能方式：

(1) 硫化氢直接与高铁氧化物反应生成，即铁获得电子被还原，硫失去电子被氧化：

$$Fe^{3+}+e=Fe^{2+};2S^{2-}-2e=[S_2]^{2-}$$

反应式：$2H_2S+Fe_2O_3=FeS_2+FeO+2H_2O$

(2) 硫化氢(或硫化物、$HS^-$)和低铁氧化物，在有游离氧参与下反应生成。即硫失去电子被氧化，失去的电子为氧所获得：

$$[O]+2e=O^{2-};2S^{2-}-2e=[S_2]^{2-}$$

反应式：$2FeO+4H_2S+O_2=2FeS_2+4H_2O$

(3) 一硫化铁被游离氧氧化，而生成 $FeS_2$。同上式一样，硫失去电子被氧化，失去的电子为游离氧所获得。

反应式： $2FeS+(O)+H_2O=FeS_2+Fe^{2+}+2OH^-$

(4) 一硫化铁与元素硫反应生成。其电子结构类似于 $S_2O_3^{2-}$。

(5) 生物化学作用生成：从最近一些资料报导，在富含有机物质的沉积岩和现代沉积物中，经常可以看到一些硫化矿物(尤其是 $FeS_2$)的显微球粒以莓状结构分布十分广泛；并通过化学处理、显微鉴定或同位素的测定证实系微生物成因。

(6) 还有一些其他形成方式的可能性。例如，在一些资料上记载着，在成岩过程中，磁铁矿经过溶蚀而转变为金红石、板钛矿、锐钛矿和黄铁矿，有利于这种转化的条件可能是还原环境。甚至有人认为，在常温常压的成岩作用过程中，也能生成黄铁矿。

总之，沉积岩中所出现的 $FeS_2$ 可以理解为：从高氧化态的 $SO_4^{2-}$(硫的氧化数为 +6)还原到低氧化态 $[S_2]^{2-}$(硫的平均氧化数为 -1)，每个硫原子不是获得 8 个电子，而是获得 7 个电子。如果与还原剂有机物质作用，那么，化学反应方程式不应该为：

$SO_4^{2-}+2C=S^{2-}+2CO_2$；

而应该为：$4SO_4^{2-}+7C+4H^+=2S_2^{2-}+7CO_2+2H_2O$

或 $4SO_4^{2-}+7C+4Me^+=2S_2^{2-}+7CO_2+2Me_2O$

(式中 $Me^+$ 代表金属离子)

因此，由方程式 $SO_4^{2-}+2C=S^{2-}+2CO_2$ 来求得“0.75”是错误的；由此相应求得的“0.91”也就是错误的。

### 2.2 菱铁矿的形成和“0.215”

齐诺维也夫的 $C_{Fe}^{2+}=0.215$ 来源于化学反应方程式：

$$2Fe_2O_3+4C+3O_2=4FeCO_3 \cdots\cdots\cdots\cdots ④$$

故 $C_{Fe}^{2+}=\frac{C}{Fe}=0.215$……生成一个重量单位的 $FeCO_3$ 的铁所消耗的有机碳的量。

我们分析一下反应方程式④就可以发现，形成 4 个分子的 $FeCO_3$，实际上只有一个碳消耗在还原 $Fe_2O_3$ 的铁上，其他 3 个碳为游离氧所氧化。现在就要提出这个反应方程式是否合理的问题。

根据氧化-还原化学反应的理论，元素的变价与否是受着它本身原子的构造和介质的氧化-还原电位控制的。一般说来，离子的氧化-还原顺序决定于氧化-还原电位的顺序性；而电极电位又决定于有关氧化剂和还原剂的相对强弱；在有几种还原剂和氧化剂同时存在的场合下，只有最强的还原剂或最强的氧化剂首先发生氧化-还原反应。在反应方程式④的反应中，有机碳作为还

原剂,高铁及游离氧作为氧化剂,它们的标准氧化-还原电位分别为:

$$2H_2O = O_2 + 4H^+ + 4e + 1.229V$$

$$Fe^{2+} = Fe^{3+} + e + 0.77V$$

$$C = C^{4+} + 4e - 0.1V$$

当然,首先应该说明,这里所引用的电位值是在标准状况下的相对值,特别是有机物质氧化的半电位之值并非完全一致和确切。更重要的是,在岩石中所进行的氧化-还原反应还受下列因素的影响:

(1) 沉积物或岩石中氧化剂或还原剂量的多少,即浓度对电位的影响。对金属离子来说(包括 $H^+$),离子浓度越大,电位值越正;相反,离子浓度越小,电位值越负。而对非金属元素来说,离子浓度越大,电位越负;离子浓度越小,电位越正。如大家所熟知的聂恩斯特公式:

$$E = E_0 + \frac{0.058}{n}\log C$$

式中:$E_0$ 为标准极电势,$C$ 为离子浓度,$n$ 为氧化-还原反应中离子平衡的电子数,0.058 为随温度变化的常数。

根据计算,当 pH = 7 时,$2H_2O = O_2 + 4H^+$ 的半电位为 0.815V;$Fe^{2+} = Fe^{3+}$之半电位值在一般情况下都可能小于 0.77V。

(2) 外部地质因素、温度、压力等的影响。如直接关系到离子在溶液中的运动速度,从而出现所谓超电势等。

(3) 氧化-还原的许多元素共存的相互影响。

(4) 受 pH 的影响。主要表现在几种铁矿物形成时,pH 的影响是相当大的,正如,克伦宾和加勒尔等以 Eh 和 pH 的关系来阐明黄铁矿-菱铁矿-赤铁矿的稳定范围那样。

更重要的是,我们还需要弄清楚,$FeCO_3$ 中的 C 应该如何对待的问题。对沉积岩来说,基本的、主要的能源(抛开地球内能和太阳的辐射能等而言)是有机物质的作用。高氧化态的元素还原到低氧化态是有机物质作用的结果;高价的铁还原到低价的铁(包括菱铁矿在内的各种形态的低铁),归根到底,都是有机物直接或间接(例如,$SO_4^{2-}$ 被有机物还原为 $H_2S$ 后,$H_2S$ 再还原高铁到低铁)作用的结果。这是无可置疑的。但是,形成 $FeCO_3$ 所需之 $CO_2$ 或 $CO_3^{2-}$ 的来源是多种多样的。除了反应方程式①以外,归纳起来,有下列几种其他的主要来源:

(1) 沉积水体中溶解的 $CO_2$:0 ℃时,$CO_2$ 在水中的溶解度(按体积计)为 1.71%,可见水体中 $CO_2$ 的含量是很可观的。

(2) $SO_4^{2-}$ 和非铁的高氧化态的金属元素等被有机物质还原所生成的

$CO_2$ 也不会太少。特别是 $SO_4^{2-}$ 等物质的含量甚高时，这个量将会是非常大的。

(3) $Fe^{3+}$、$SO_4^{2-}$ 还原之前的整个沉积、成岩作用过程中，由于有机物质和游离氧的氧化作用直至造成还原环境所生成的 $CO_2$ 量更多。对此，石油地质工笔者曾评论过，它和甲烷等一样是无法估计的。

(4) 水体和沉积物中其他形式的 $CO_3^{2-}$、$HCO_3^-$ 的分解(或取代反应)都会提供出形成 $FeCO_3$ 所需的 $CO_2$ 和 $CO_3^{2-}$。如沉积物下沉或成岩时，可以直接为其他碳酸盐所替代。

(5) 有机物的去羧基作用以及生物作用所生成的 $CO_2$ 也是不可忽视的。

(6) 其他地点(左右隣区或先沉积物)的高氧化态的金属元素被有机物还原和 $SO_4^{2-}$ 的被有机物还原所生成的 $CO_2$ 自达平衡后所剩余的部分游离而来，也将参与生成 $FeCO_3$ 的反应。

综合上述种种来源，就可以得出结论。$FeCO_3$ 形成时所需的 $CO_2$ 远远不像反应方程式④那样由于大量游离氧参与下同时消耗大量的有机物质所生成的。这就是反应方程式④和 $C_{Fe}^{2+}=0.215$ 存在着的严重错误的一个重要方面。

同时，从氧化-还原反应的观点及其计算方法来看，$C_{Fe}^{2+}$ 的计算没有正确反映氧化-还原反应中的式量和当量之间的关系，因而，也就无法反映出铁和有机碳之间的质量关系。显然，不能用这种不正确的关系来进行氧化-还原反应的定量计算。

另外，历来在分析工作和成果的计算中，$Fe^{2+}$ 是以 5% HCl 可溶性的铁笼统地来应用，并将 $Fe^{2+}$ 全部作为 $FeCO_3$ 的形式存在来对待的。实际上，沉积岩中，还广泛地与 $FeCO_3$ 共生的有低铁型的氧化物和硅酸盐，如磁铁矿、氢氧化亚铁、绿泥石等。对于这部分类型的低铁化合物，如果同样以方程式④来代替(何况前已述及，方程式④实际上是不存在的)就同样存在着明显的片面性。

因此，上述种种理由表明，不能应用反应方程式④来定量计算 $C_{Fe}^{2+}$ 值，不应该采用“0.215”这个系数。

无须再述，既然“0.91”和“0.215”都是错误的，那么，“0.236”也就是错误的了。

## 3 新的铁的还原系数($x_{Fe}$)较为合理

上面批判了齐诺维也夫所引用的几个化学反应方程式以及“0.91”、

“0.215”和相应计算出来的“0.236”，那么，正确的反应方程式和合理的系数是什么？

石油有机成因的近代理论表明，富含氧的原始有机物质转变为贫氧的石油是处在一种特定的介质中，而显示出原始有机物质强烈地被还原的结果；石油的生成必须具备一个基本的外部因素——还原环境。原始有机物质转变为石油的一系列过程本身就是一种氧化-还原的变化过程，环境的演变过程也就是一个氧化-还原变化的过程。

众所周知，氧化-还原反应的实质是一种元素或化合物失去电子而另一种元素或化合物获得（或吸引）电子所组成的反应。因此，用电化学的理论，用电子得失或转移，或电子之密度的增大和减小，或称之为极性数的改变的方法，来解决 $FeCO_3$ 和 $FeS_2$ 的形成及其相互关系，从而把这种关系应用到铁的还原系数中去就是最为合理的。这种矛盾的普遍性无疑寓于这样的特殊性之中。新的铁的还原系数就是在这种理论和思想指导下推导出来的。下面用几个方法来阐明：

3.1 用电子平衡的方法

① $Fe^{3+} = Fe^{2+} - e$

② $2S^{+6} = S_2^{2-} - 14e$

在离子电子方程式中：三价铁获得 1 个电子被还原到二价；两个硫酸盐中的硫获得 14 个电子被还原为正$[S_2]^{2-}$。在 $FeS_2$ 分子中，有 1 个铁原子和 2 个硫原子，铁从三价还原到二价，硫从正六价还原到“负一价”（平均数），总共获得 15 个电子。而在 $FeCO_3$ 中，仅有 1 个铁原子，就只获得 1 个电子。由此可见，形成 $FeCO_3$ 和形成 $FeS_2$ 所需的电子数之比为立 1 ∶ 15（或 $\frac{1}{15} = 0.0667$）。

3.2 用有机碳还原的平衡计算法

反应式：① $4SO_4^{2-} + 7C + 4Me^+ = 2S_2^{2-} + 7CO_2 + 4MeO$

② $4Fe^{3+} + C = 4Fe^{2+} + C^{4+}$

该组反应式表明，1 个 $S_2^{2-}$ 相当于 3.5 个碳，1 个铁相当于 0.25 个碳。

所以 $\frac{FeCO_3}{FeS_2} = \frac{0.25}{0.25 + 3.5} = \frac{0.25}{3.75} = \frac{1}{15} = 0.0667$。

3.3 用氧化剂氧化反推法

反应式：

$$①\begin{cases}4FeS_2+15O_2=8SO_3+2Fe_2O_3\\4FeCO_3+O_2=4CO_2+2Fe_2O_3\end{cases}$$

$$②\begin{cases}2FeS_2+5K_2Cr_2O_7+19H_2SO_4=5K_2SO_4+5Cr_2(SO_4)_3Fe_2(SO_4)_3+19H_2O\\6FeCO_3+K_2Cr_2O_7+13H_2SO_4=K_2SO_4+Cr_2(SO_4)_3+3Fe_2(SO_4)_3+\\\qquad 13H_2O+6CO_2\end{cases}$$

从第①组反应式来看,4 分子的 $FeS_2$ 需 15 分子氧;而 4 分子的 $FeCO_3$ 仅需 1 分子氧。所以,其比值为 1 : 15。从第②组反应式来看,1 分子的 $FeS_2$ 相当于$\frac{5}{2}$分子的 $K_2Cr_2O_7$;而 1 分子的 $FeCO_3$ 相当于$\frac{1}{6}$分子的 $K_2Cr_2O_7$。其比值仍为 1 : 15,

即$\frac{FeCO_3}{FeS_2}=\frac{\frac{1}{6}}{\frac{5}{2}}=\frac{1}{15}=0.0667$。

### 3.4 用化学分析方法

试验:准确称取 $FeCO_3$ 和 $FeS_2$ 的单矿物样品,分别用 $K_2Cr_2O_7\cdot H_2SO_4$ 溶液进行氧化,反应完全后,用莫氏盐溶液滴定之。由此消耗的氧化剂 $K_2Cr_2O_7$ 的量分别计算出两种矿物对 $K_2Cr_2O_7$ 消耗量的比值,其数据列表如下:

| 项目 \ 数据 \ 矿名 | $FeCO_3$ | $FeS_2$ | 备注 |
|---|---|---|---|
| 样品质量(g) | 0.5 | 0.1 | |
| 加入 $K_2Cr_2O_7$ 量(mL) | 15 | 45 | |
| 实际消耗 $K_2Cr_2O_7$ 量(mL) | 7.8 | 26.77 | $FeCO_3$ 纯度为 74.15%,$FeS_2$ 纯度为 89.8% |
| 0.1 g 分子消耗量(mL) | 243.4 | 3576 | |
| 比值 | 1 | 14.68 | |

由表可以看出:对 0.1 mol 的 $FeCO_3$ 和 $FeS_2$ 来说,分别消耗 $K_2Cr_2O_7$ 的量为 243.4 mL 和 3576 mL,其比值为$\frac{1}{14.68}=\frac{FeCO_3}{FeS_2}$。其所以为 14.68,而不为 15,这是由多方面原因所造成的,主要是实验中的误差(可看出部分 $FeS_2$ 未分解完全)。

上述 4 种方法清楚地表明,在 $FeCO_3$ 和 $FeS_2$ 形成的氧化-还原变化中,以氧化-还原理论进行指导,并初步经化学分析得到验证,两者之间量的关系是一定的,同样均为 1 : 15。

因此,新的铁的还原系数(称之为 $x_{Fe}$)采用的方程式则是:

(a) $4SO_4^{2-} + 7C + 4Me^+ \longrightarrow S_2^{2-} + 7CO_2 + 2MeO$

(b) $2Fe_2O_3 + C \longrightarrow FeCO_3 + 3FeO$

即形成 $FeCO_3$ 的计算采用反应式(b);形成 $FeS_2$ 的计算采用反应式(a)与(b)之和。

其计算式则为:

$$x_{Fe} = \frac{Fe^{2+}_{HCl} + Fe^{2+}_{FeS_2} \times 15}{\Sigma Fe}$$

或者 $x_{Fe} = \dfrac{Fe^{2+}_{HCl} \times 0.0667 + Fe^{2+}_{FeS_2}}{\Sigma Fe}$

式中:$Fe^{2+}_{HCl}$——稀盐酸溶解的低铁总含量;

$Fe^{2+}_{FeS_2}$——$FeS_2$ 形态的铁的含量;

$\Sigma Fe$——稀盐酸溶解的高、低铁含量与 $FeS_2$ 形态的铁的含量的总和。

$x_{Fe}$前一种形式和后一种形式在理论上和实质上没有区别,都可以应用。但是,在实际应用当中,前者可能比后者好,主要表现在两个方面:

① 在地质上应用时,可以略去不必要的多余的有效数字位数,即在大多数情况下,可以免去小数点前后的一个零或两个零。同时,在计算时,乘 15 比乘 0.0667 要方便一些。

② 在化学分析上,提出了对还原硫测定的更高精度的要求,这也是有利于指标的应用的。

## 4 评论

*K* 值或者齐诺维也夫、古柳也娃等仅仅在形式上把铁的自生矿物的形成视为氧化-还原反应,而没有抓住氧化-还原反应的实质。当然,不可否认,随同沉积物的水体或介质中游离氧一定也会消耗有机物质,其他变价元素如 Mn、V 等也会与沉积介质发生电子得失、交换、转移的反应,但是,除了已经阐述的错误外,他们没有清楚地意识到,我们的指标是铁的还原系数,是对铁的系统来说的,不同的矛盾只有用不同的方法来解决。

电子的得失或电子高密度的增减,是氧化-还原反应与其他化学反应的特殊性,对于氧化-还原反应系统中的任何一个反应来说,则具有共同的普遍的规律性。因此,我们在研究氧化-还原这一类型的反应时,就必须用这种普遍的规律性来指导,并遵循这一事物发展的总的规律。

在沉积作用(实际上可以认为时间是很短暂的)和成岩作用这一个总的

系统的物质运动发展长途中,我们还可以分析一下发展过程的阶段性,就可以对 $x_{Fe}$ 有更为深刻的了解。这也是齐诺维也夫等所没有注意的问题,即事物发展的整个过程中存在着阶段性,不同的阶段有不同的特点。沉积作用和成岩作用这个过程对有机物质来说,根据氧化-还原反应,可以分为两个各具特点的阶段:

(1) 有游离氧的氧化-还原反应阶段,也可称为除游离氧的阶段。这个阶段的作用主要是有机物质与游离氧的作用。

(2) 无游离氧的氧化-还原反应阶段,也可称为形成还原矿物的阶段。这个阶段的作用主要是有机物质与高价金属元素化合物或与 $SO_4^{2-}$ 等直接或间接的作用,推测其中应该认为伴随着有机物质转化的反应。

如果第一阶段的反应没有进行,或者没有进行完毕,就不会有第二阶段的反应发生。这就是说,游离氧不消耗掉就不会有还原矿物的生成。假如介质中游离氧十分丰富,将有机物全部氧化完后还有剩余,或者沉积物中有机物含量很少,全部消耗在第一阶段就不能进入第二阶段,那么,这种介质就是典型的氧化环境。消耗在第一阶段的有机物质的量,如果水不流动,或处于水体深处、波浪影响不到的区域,或者沉积物堆积非常迅速、处于“闭塞”状态,那么,所需要的有机物质的量主要是与溶于水中游离氧氧化反应的量。反之,如果水体流动,沉积物不是处于“闭塞”状态,则这种消耗就会随着水体活动的增强而增加,或随着地质构造运动的增强而增加。但是,事实上,我们很难估计游离氧的量,也不必要估计第一阶段所消耗的有机物质的量。

因此,第二阶段才形成 $FeCO_3$,这时的 $FeCO_3$ 就不应该考虑有游离氧的参与。据此也可以进一步阐述前述的反应方程式①由于时间和地点的差别而不应该运用这个方程式的原因。对于 $FeS_2$ 来说,虽然也可理解为分两步进行,但它的形成则处在同一阶段之中。当然,这是一般的规律。在个别特殊的情况下,比如当 $pH>9$、铁离子浓度很大时,就有可能有机物先与铁反应,而后与游离氧反应;或者二者几乎同时进行;或者二者或先或后地交换进行。但是,在有机物质较为丰富的情况下,应该看到,由于它的氧化所形成的有机酸和 $CO_2$ 都会大大降低介质的 pH 值。这也就是石油地质工笔者把 $FeCO_3$ 这种指相矿物可作为氧化相,而在大多数情况下可作为还原相的矿物来对待的理论依据。

总之,通过多方面的探讨,人们完全可以清楚地看到,$K_1$ 值(即“0.25”)错误明显,这是无可非议的。对比 $x_{Fe}$ 系数与 $K$ 系数,无疑 $x_{Fe}$ 系数具有更为完善的和坚实的理论基础。从而可以认为,$x_{Fe}$ 是所有铁的还原系数中最为合理的系数。

## 5 结束语

全篇文章基本上都是理论上的阐述，带有浓厚的假说性，并未涉及地质应用，也未提出划分沉积地球化学相的具体方法及其标准。可以预料，变化多端的客观地质体往往比一般由理论研究所知道的复杂得多。因而，可以说没有得到地质实践的验证，的确还只能说仅仅认识了对于事物全貌认识的并非十分重要的一半，还有更为重要的一半没有得到解决。

这篇文章经孙肇才、陈克仁、江维林、刘永洲、李令德、王犧、沈志根、倪径明、张子培、张玉琴、刘秀菊、吴文等过目或讨论，特别是何中恒、王中学、张洪林等提出了很多宝贵意见，特作说明，并表示感谢。

### 参考文献

[1] 地质部石油地质局中心实验室. 石油地质实验，1964(6).

[2] 成都地质学院. 沉积相及古地理教程[M]. 北京：中国工业出版社，1961.

[3] 南京大学地质系. 地球化学，1964.

[4] 维也纳茨基 B. И. 地球化学概论[M]. 北京：科学出版社，1962.

[5] 古柳也娃 Л. A. (гуляева). 沉积岩的地球化学相、氧化-还原环境及其中的有机物[M]. //矿产专辑(第5辑)，北京：地质出版社，1957.

[6] 斯特拉霍夫 HM. 沉积物之成岩作用及其对沉积矿物形成的意义[M]//沉积专辑(第1辑)，北京：地质出版社，1958.

[7] 詹姆斯 HL. 含铁建造沉积相[M]. 北京：中国工业出版社，1962.

[8] 格里戈里也夫 C. M. 可燃矿产形成过程及性质[M]. 北京：地质出版社，1959.

[9] 成都地质学院. 结晶学及矿物学[M]. 北京：中国工业出版社，1961.

[10] юоркевич ИА, исследования по методике фацильно-геолимического изучения осадочных пород, 1958.

[11] Геохимического исследования (перевод с английского) сборник статей, издательство иностранной литературы, 1961.

本文原载于《石油实验地质》1965 年第 1 期

# 19．红外光谱在石油地质研究中的几个问题

# Infrared spectroscopy applications in petroleum geology few questions

何志高

红外光谱引入我们的石油地质领域已有 15 年的历史。以往主要进行了大量地质应用方面的研究，而基本理论和方法的研究却不多。在地质应用研究中，以“芳烃结构分布指数”应用最普遍。本文重点拟通过对“芳烃结构分布指数”的剖析，提出一些具有普遍意义的认识。很多试验系与浙江石油队、荆门炼油厂研究所的同人共同完成的。

## 1　关于芳烃结构分布指数

E. E. 布雷通过大量样品的红外光谱研究，提出了一个识别和确定油源岩的指标，用以从事地球化学勘探。该指标即为芳烃结构分布指数，指标数值以 12.4 μ 与 13.4 μ 两个红外光谱吸收峰的吸收强度比值表示，即 A12.4 μ/A13.4 μ；指示油源岩的数值界限为 0.8～1.4。布雷认为，该指标也适用于碳酸盐岩，比正烷烃的碳数奇偶优势指标应用更为广泛。我们应用多年，多数实践效果不好。

根据最近的研究，应用效果不好的原因主要是由于分析方法不善所造成的。而且更重要的是，以往我们对于该指标理论上的意义认识不足，指标应用范围及条件认识不足，尚有不少错误的理解。此外，在布雷对指标的12.4 μ与13.4 μ的光谱归属问题上以及样品处理方法问题上，本文也提出一些不同意见，一一阐述如下，供讨论参考。

### 1.1　关于苯环上的氢原子的概念问题

苯类化合物有一个重要化学性质，就是环上的氢原子不稳定，容易发生取代反应，所以，通常把这种氢原子称为初级氢原子或自由氢原子。由表 1 可以清楚地看出，这类氢原子有“相互邻近”与“不相互邻近”之分。取代基数量

与对应的相互邻近的初级氢原子的数量是两个不同的概念，不能混淆这些概念。布雷的两类氢原子都是相互之间邻近的，这种“邻近”不是氢原子与碳原子的邻近，而是氢原子之间的邻近。以二甲苯为例，3 种异构体都是 2 个取代基，4 个初级氢原子。但是，相互邻近者却有 3 种类型。1，2-二取代为一种类型，4 个氢原子相互邻近；1，3-二取代为两种类型，一种是单独的一个，另一种是三个相互邻近；1，3-二取代也有两种类型，但二者都是 2 个相互邻近。这个概念布雷是讲清楚了的，是一个最基本的必须搞清楚的概念。

1.2　12.4 μ 与 13.4 μ 的光谱归属问题

12.4 μ（相当于 806 $cm^{-1}$）：布雷归属于芳核上 1 ~ 2 个相互邻近的初级氢原子，称之为第Ⅰ类氢原子；

13.4 μ（相当于 746 $cm^{-1}$）：布雷归属为 4 ~ 5 个相互邻近的初级氢原子，称之为第Ⅱ类氢原子。

我们认为，12.4 μ 的红外光谱归属不是芳核上 1 ~ 2 个相互邻近的初级氢原子，而是 2 个相互邻近的初级氢原子。初级氢原子非“2”个不可，“1”个是不行的，“1”个不能列入此类。苯环上或稠环芳烃上处于“对位”的 2 个初级氢原子不是相互邻近，不能称为“相互邻近”，它们不能列入Ⅰ类氢原子范畴。

表 1 是单环芳烃结构分布与对应的红外光谱特征，表 2 是与石油有关的稠环芳烃结构及其对应红外光谱特征。这两张表清楚地给出了上述结论的充分依据。例如蒽和菲，在布雷的芳烃结构概念中，都是 2 个Ⅰ类氢原子，8 个Ⅱ类氢原子，按此道理，Ⅰ类氢原子就应该都有 12.4 μ（或 806 $cm^{-1}$）的吸收谱带。但是，事实并不是这样，蒽的红外光谱图上根本就没有 12.4 μ（806 $cm^{-1}$）的吸收带，如图 1 所示。

从图 1 可见，菲的确存在着两个相互邻近的初级氢原子，光谱特征正好反映了结构上的这种特征，在 746 $cm^{-1}$ 处有一个吸收带，归属于 4 个相互邻近的初级氢原子，在 806 $cm^{-1}$ 处有一个吸收带，这就是两个相互邻近的初级氢原子的特征谱带，而蒽的中间苯环上的“对位”两个初级氢原子不是“相互邻近”，因而，12.4 μ（806 $cm^{-1}$）处毫无吸收。这种蒽类的两个初级氢原子在结构上、光学性质上、化学性质上均与菲当中苯环上的两个相邻初级氢原子存在着明显的差别，这是由分子中电荷分布能量差异造成的，可以说是蒽与菲的质的区别所在。

为了说明表 2 所示的与石油有关的各种稠环芳烃也都可以用上述理论进行解释，求得其理论上的具有普遍性意义的认识，以图 2、图 3 的稠环芳烃为例，剖析其结构特征与光谱特征之间的关系。

**表 1 单环芳烃结构分布与红外光谱特征**

| 类型与数量 | 初级氢原子总数 | 相邻氢原子数 | | 红外光谱特征波数($cm^{-1}$) |
|---|---|---|---|---|
| | | a 类 | b 类 | |
| 单取代 | 5 | 5 | | 770~730,710~690 |
| 1,2-二取代 | 4 | 4 | | 770~735 |
| 1,3-二取代 | 4 | 1 | 3 | 810~750,710,690 |
| 1,4-二取代 | 4 | 2 | 2 | 833~810 |
| 1,2,3-三取代 | 3 | 3 | | 780~760,740~705 |
| 1,2,4-三取代 | 3 | 1 | 2 | 825~805,885~870 |
| 1,2,5-三取代 | 3 | 1 | 2 | 860~810,730~765 |
| 1,2,3,4-四取代 | 2 | 2 | | 810~800 |
| 1,2,3,5-四取代 | 2 | 1 | 1 | 850~840 |
| 1,2,4,5-四取代 | 2 | 1 | 1 | 870~850 |
| 五取代 | 1 | 1 | | 870 |

**表 2 与石油有关稠环芳烃结构及其对应的红外光谱特征**

| 结构 | | 取代类型(对应单环) | 光谱特征 | | 相邻氢原子数 | |
|---|---|---|---|---|---|---|
| 结构式 | 单元 | | a | b | a | b |
| 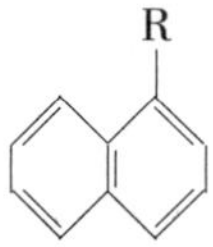 | α | 1,2-二取代 | 770~735 | | 4 | |
| | β | 1,2,3-三取代 | 780~760 | | 3 | |
| 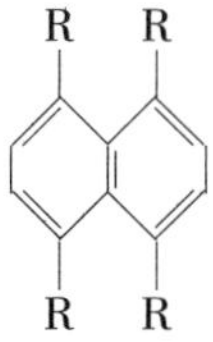 | α | 1,2,3,4-四取代 | 810~800 | 740~705 | 2 | |
| | β | 1,2,3,4-四取代 | 810~800 | | 2 | |
| 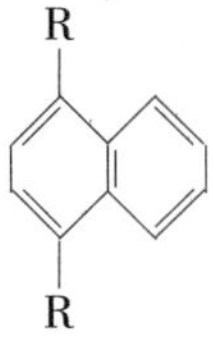 | α | 1,2,3,4-四取代 | 810~800 | | 2 | |
| | β | 1,2-二取代 | 770~735 | | 4 | |

续表

| 结构 | | 取代类型（对应单环） | 光谱特征 | | 相邻氢原子数 | |
|---|---|---|---|---|---|---|
| 结构式 | 单元 | | a | b | a | b |
| 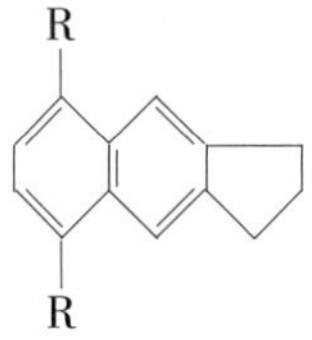 | α | 1,2,3,4-四取代 | 810～800 | | 2 | |
| | β | 1,2,4,5-四取代 | 870～850 | | 1 | 1 |
| 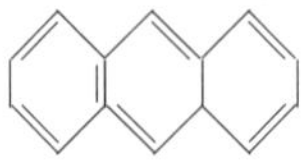 | α | 1,2-二取代 | 770～735 | | 4 | |
| | β | 1,2,4,5-四取代 | 870～850 | | 1 | 1 |
| | γ | 1,2-二取代 | 770～735 | | 4 | |
| 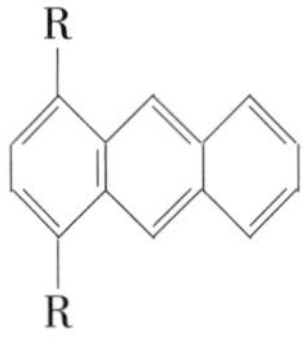 | α | 1,2,3,4-四取代 | 810～900 | | 2 | |
| | β | 1,2,4,5-四取代 | 870～850 | | 1 | 1 |
| | γ | 1,2-二取代 | 770～735 | | 4 | |
| 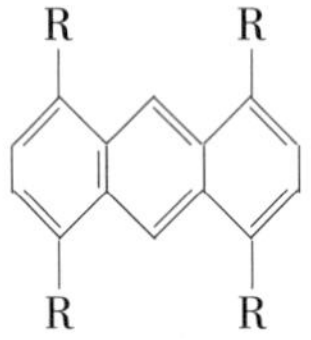 | α | 1,2,3,4-四取代 | 810～800 | | 2 | |
| | β | 1,2,4,5-四取代 | 870～850 | | 1 | 1 |
| | γ | 1,2,3,4-四取代 | 810～800 | | 2 | |
| 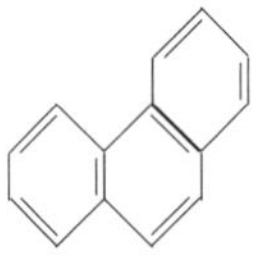 | α | 1,2-二取代 | 770～735 | | 4 | |
| | β | 1,2,3,4-四取代 | 810～800 | | 2 | |
| | γ | 1,2-二取代 | 770～735 | | 4 | |
| 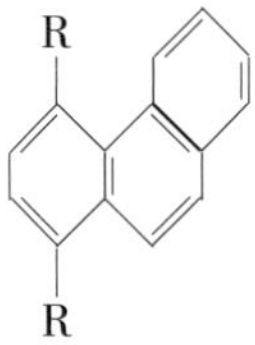 | α | 1,2,3,4-四取代 | 810～800 | | 2 | |
| | β | 1,2,3,4-四取代 | 810～800 | | 2 | |
| | γ | 1,2-二取代 | 770～735 | | 4 | |
| 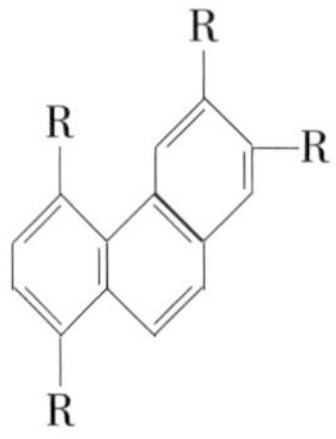 | α | 1,2,3,4-四取代 | 810～800 | | 2 | |
| | β | 1,2,3,4-四取代 | 810～800 | | 2 | |
| | γ | 1,2,4,5-四取代 | 870～850 | | 1 | 1 |

续表

| 结构 | | 取代类型(对应单环) | 光谱特征 | | 相邻氢原子数 | |
|---|---|---|---|---|---|---|
| 结构式 | 单元 | | a | b | a | b |
| 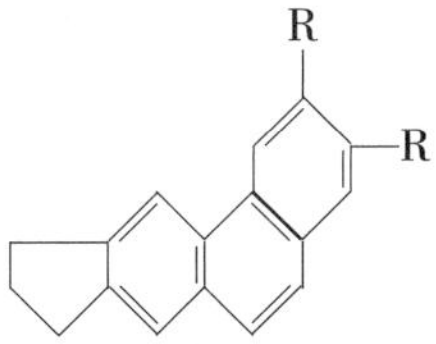 | α | 1,2,4,5 -四取代 | 870 ~ 850 | | 1 | 1 |
| | β | 1,2,3,4 -四取代 | 810 ~ 800 | | 2 | |
| | γ | 1,2,4,5 -四取代 | 870 ~ 850 | | 1 | 1 |

注:结构式中从左至右的芳环结构单元分别用 α、β、γ 代表。

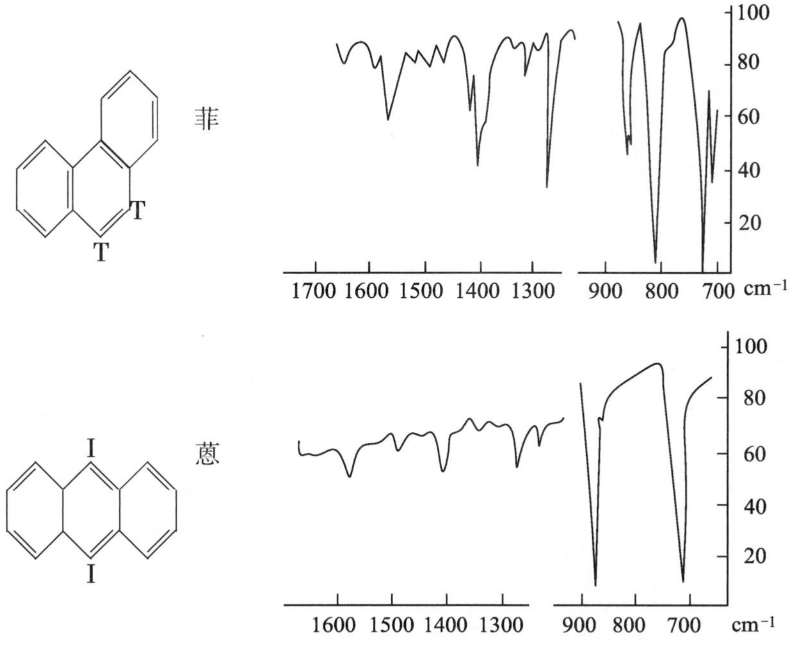


**图 1 蒽、菲结构与光谱对比(引自 DMS)**

图 2 是一个具有 R、R′取代基的双环芳烃。分子的左边环可以认为相当于单环芳烃的 1,2,3,4 -四取代基的结构,β-位置上的 2 个初级氢原子相互邻近,对应的光谱特征吸收是 810 ~ 800 $cm^{-1}$;分子右边环可以认为相当于单环芳烃的 1,2 -二取代的结构,环上 4 个氢原子相互邻近,对应的光谱特征吸收是 770 ~ 735 $cm^{-1}$。图 3 是一个具有环烷烃的芳烃,该环环烷环上具 R、R′、R″取代基。对于这样的分子,其分子的左边环相当于 1,2 -二取代的单环芳烃,4 个初级氢原子相互邻近,对应的光谱特征吸收为 770 ~ 735 $cm^{-1}$;分子的右边

环相当于1,2,4,5-四取代的单环芳烃,对应的光谱特征吸收由表1可看出为870~850 $cm^{-1}$。

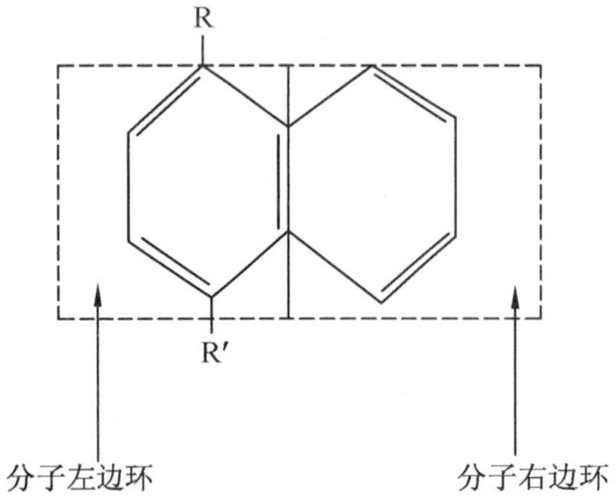


**图2　具有R、R取代基的双环芳烃**

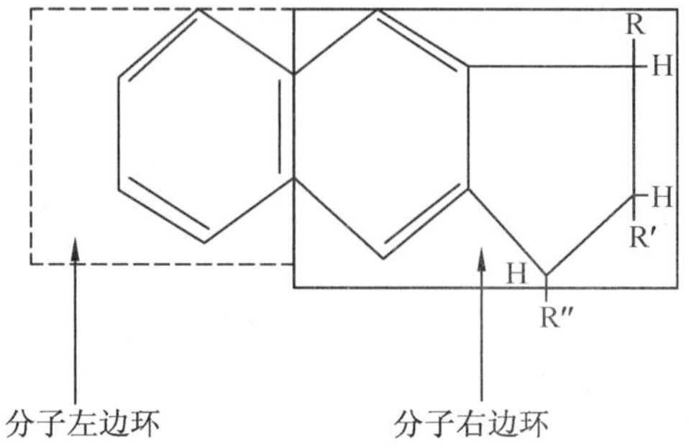


**图3　具有环烷烃的芳烃**

理论和实际是完全吻合的。稠环芳烃均可以应用单环芳烃的取代类型解释光谱的归属,频率位移的问题,对双环、三环影响很小,而在石油类芳烃中,除去单环以外,双、三环芳烃可占绝大部分。

以上是12.4 μ(806 $cm^{-1}$)的归属问题,对于13.4 μ的归属也有值得讨论之处。布雷将13.4 μ归属于4~5个相互邻近的初级氢原子。我们认为,这种归属对于单环芳烃来说是可以的,从表1可以看出,单取代的单环芳烃对应于5个相互邻近的初级氢原子,光谱特征有两个吸收带,其中一个即为~746 $cm^{-1}$。这个归属对联苯来说也是可以的,但是对稠环芳烃来说则是不可以的。因为在稠环芳烃中没有也不可能有5个相互邻近初级氢原子结构的化合物或者结构体系,也可以说,5个相互邻近的初级氢原子在任何稠环芳烃中都不存在。同时,鉴于样品经历去除轻芳烃的处理后,样品中实际上已不存在只有一个单取代的单环芳烃。即使有,含量也是很低的,不可能对光谱吸收带造成影响。对联苯来说亦如此,对于一个单取代形式的环烷环的烃类来说亦如此(当然主要是讲含量低)。因此,在确定13.4 μ的归属时,应该定为4个相邻初级氢原子比较合适。

总之,12.4 μ和13.4 μ的归属问题是一个地球化学上的严肃的问题,也是有机化学、光谱学上的一个不可含糊的问题,在我们的研究中是一个可以引导深入的问题。13.4 μ的归属区问题较小,而12.4 μ的归属就不可忽视了。

### 1.3　芳烃结构分布指数的变化

按布雷本意,芳烃结构分布指数是作为确定油源岩、划分油源岩与非油源岩的指示标志。凡12.4 μ与13.4 μ吸收强度比值位于0.8~1.4范围者称油源岩，否则就是非油源岩。（当然也规定了例外的情况。）

在我们的实际应用中，提出来了一些问题：随着演化，比值变化的规律是什么？是否比值越大越好？例如，有的资料说，芳烃进一步缩合时，芳核侧链上4～5个邻近氢原子数增加；在划分“成煤区”“成油区”的坐标图上似乎比值越大越好，在“成油区”比值可以达到远远大于1.4的程度。

最近研究也有些新的认识，是否应该这样认识？对比图4可以看出，芳核侧链上4个相邻初级氢原子随缩合程度变化而有增有减。例如，当芳核处在断支链阶段或者断支链以后，是增加的；而当芳核侧链上支链很少，芳核进一步缩合时，4个相邻初级氢原子将减少，甚至达到像图4中的七环芳烃，4个相邻氢原子全不存在，全部都是2个相邻初级氢原子。同时，这种由大变小的规律导致了12.4 μ(806 $cm^{-1}$)的吸收增强，13.4 μ(746 $cm^{-1}$)的吸收降低，其12.4 μ/13.4 μ或806 $cm^{-1}$/746 $cm^{-1}$增大。如果是这种情况，当然比值不是越大越好。因而，坐标上的“成油区”是否应该将此比值确定于一个最大上限值才比较合理。

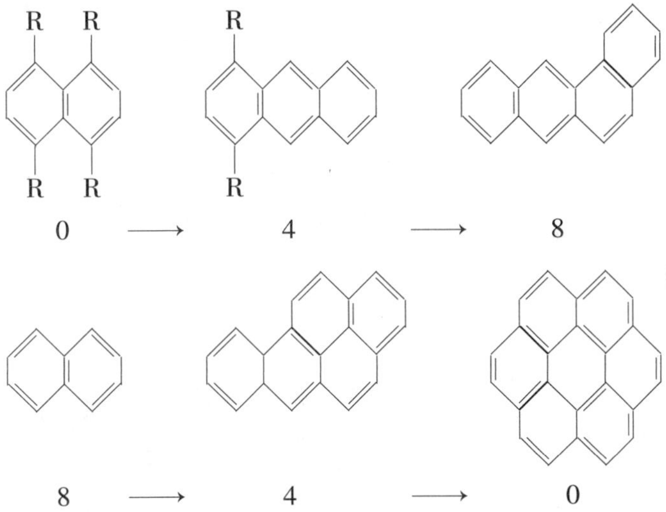


**图4　4个相邻初级氢原子数的变化**

另外，结构分布指数数值的大小从理论上来看不会是如布雷所说，由两类“氢原子的丰度”所决定。因为，从各类单化合物的标准图谱来看，如表1中各类取代基苯，以对应的光谱吸收带的吸收强度作比较，例如具5个相互邻近的初级氢原子的单取代苯与具单独一个初级氢原子的五取代基苯作比较，面外弯曲振动特征峰的吸收强度相差不多，并不存在“氢原子丰度”的比例关系。因此，只能说指数数值大小或吸收带的吸收强度主要取决于芳烃分子中每一个具一定数量氢原子的“苯结构单元的丰度”以及对应于该吸收带的归属结构的偶极矩。关于吸收强度的问题，本文后面还要述及，此处不作更多的讨论。

### 1.4 芳烃结构的显示

芳烃结构分布指数最基本的要求是把“芳烃结构”充分而又真实地显示出来。为此，应该有两条措施：一是要有足够的芳烃样品浓度；二是要消除各类干扰。

#### 1.4.1 关于浓度（或者厚度）

这是显示结构分布的最基本的条件。浓度低或者样品薄，该区域吸收很弱，806 $cm^{-1}$、746 $cm^{-1}$强度就会很低，测量困难、测量误差也大，不符合比尔定律的定量分析区域。我们统计了过去红外光谱的分析资料，在大量的图谱中，806 $cm^{-1}$、746 $cm^{-1}$的表观吸收度（过去称光密度）在0.1以下者占全部图谱的40%以上。这部分图谱如果以2000 $cm^{-1}$吸收度作为“统一空白”，则有很大部分无法减差；如果以“切绒法”计算，其吸收度仅为0.02左右。相对于如此低的吸收度，测量误差可达8%，仪器误差（通常以仪器允许透过率误差0.5%～1%计）可达10%～20%。假若样品加厚，表观吸收度增加到0.3，那么，上述测量误差与仪器误差之和可降低到3%以下，准确度可提高5～10倍。

至于样品厚度增大以后其他吸收峰出现“打顶”的问题，可以不使用打峰顶。更好的办法是采用不同厚度分段扫描。事实上，这种方法早已被人们普遍采用，在DMS图谱集中，一个化合物的中红外区的扫描有4种不同浓度。布雷也是仅在10～15 μm范围内扫描。

#### 1.4.2 关于消除干扰的问题

单环轻芳烃类一定要去掉。因为石油类芳烃中，各种取代苯类化合物是广泛分布的，而且含量不低。从表1可以看到，700～900 $cm^{-1}$波段，各种取代苯有很多特征吸收，而且吸收强度、吸光系数都是相当大的。因此，要去掉这部分烃类。但是，要注意防止氧化。另外，饱和烃、非烃、沥青质都有干扰，所以样品要纯。

### 1.5 关于芳烃结构分布指数地球化学意义的探讨

石油中主要芳烃类化合物大致可分为以下三类：

（1）单环芳烃，通式为 $C_nH_{2n-6}$。此系列不属讨论之列。

（2）多环芳烃，通式为 $C_nH_{2n-p}$（$P=12$、14、18、20、24、30），其中 $P=14$、$P=20$之类的芳烃都是指带有一个环烷环的芳烃，都是稠环芳烃。

（3）环烷芳烃，通式为 $C_nH_{2n-m}$（$m=8$、10、12、14、16、18、20）。这类化合

物均含有一个以上的芳烃环。芳烃环也是稠环。

石油中单环芳烃类化合物含量很高，一般可占总芳烃的50%左右，这类烃由于沸点较低，大部分低于280 ℃，恒重过程（以常压、70 ℃为例），280 ℃以前的馏分没有了。所以，用作分析的样品，实际上单环芳烃很少。在多环化合物中，主要是双环系列化合物，其次是三环类化合物，高于三环类化合物较少，而且随环数增加而递减。双环芳烃的量一般平均都为三环结构量的2倍或2倍以上。

上述几种芳烃数量上的分布特征以及双环和三环上的更加独特的结构特征是石油中所独特具备的。

作为生油岩，虽然有多种定义，但是，最基本的一条是该岩石生成了并且含有石油类物质（除了受强烈变质作用外）。否则，就不能称得上"生油岩"。因此，如果岩石中含有石油所独特具备的芳烃系列化合物的含量比例关系与分子结构分布关系，那么，这种岩石就可以认为是生油岩或者具备生油岩最基本的条件。

其实，芳烃结构分布指数作为生油岩指标，与其他指标一样，有着上述这种共同的对生油岩定义的基本依据，可以说是地球化学工作一脉相承的指导思想。正烷烃奇偶优势值的降低主要是由于无明显奇偶优势的石油正烷烃在不断地大量生成后，混进原来岩石已有的正烷烃奇数优势之系统中，成为平均化的或者是被"稀释"了的奇偶优势值。石油生成量越多，降低越多，其奇偶优势值也就越接近1，接近1就认为已"成熟"，只要具有外界条件和岩石的内部条件，就可以作为油源岩向外轨送石油了。当然，如果岩石中原始生物成因的部分含量很高，"稀释"作用很弱，降低效应不明显，容当另作别论。对于芳烃结构分布指数来说，也具有与此相类似的情况，这也是任何生油指标的共同依据。无疑，对于芳烃结构分布指数来说也存在着这种"平均化"和"稀释"的过程。只不过重芳烃相对正烷烃来说，被岩石吸附的能力较强，沸点较高，从生油岩向外运移扩散的能力较低，较少受生物的降解作用。但是，布雷也提出了限定的条件，如岩石样品中有机质含量不应该超过6%，对碳酸盐岩最好不超过1%，岩石的渗透率不应该大于1 mD。

上述芳烃含量和分布在石油中的独特性，正好能清楚地、比较准确地反映在红外光谱的芳烃面外弯曲振动区，从而构成了芳烃结构分布指数810 $cm^{-1}$/750 $cm^{-1}$及其所代表的地球化学意义。图5是一种典型的原油重芳烃红外光谱图中700～900 $cm^{-1}$区域的曲线部分。

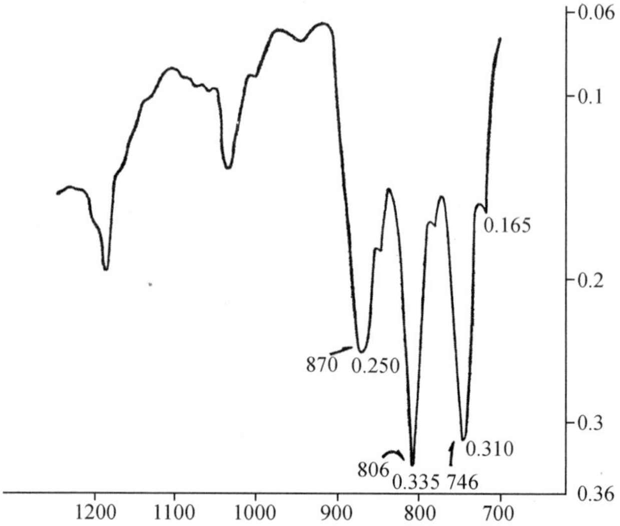


**图 5 原油重芳烃红外光谱图**

这种实例是上述理论的有力检验。图 5 上的 870 $cm^{-1}$、810 $cm^{-1}$、750 $cm^{-1}$三条吸收谱带，吸收强度相近，谱带宽度相近，谱带间距相近，810 $cm^{-1}$、750 $cm^{-1}$吸收峰锐度也相近，形状也相似。这些特征都是石油重芳烃所独特具备的特征，从外表观之，一目了然。当然，对芳烃结构分布指数的认识还很不深刻，分析方法包括化学处理与仪器分析都有不少问题，地球化学意义及其机制还不很清楚，所以，也还不能认为是很成熟的指标。

## 2 从光谱特征上讨论其他指标

自从红外光谱应用以来，许多人提出并使用了大量的指标，有的在一张图谱上甚至选取了七八个吸收峰比值作为指标，以求从中找出某些石油地质规律，同时，还相应提出了光谱中各吸收峰的归属。我们认为，其中都有不少问题是需要讨论的。但是，这里讨论的重点是从光谱学和光谱技术上来谈有关的而且是主要的指标。

### 2.1 关于 1600 $cm^{-1}$ 与 1460 $cm^{-1}$

首先应该注意，一个吸收带（峰）的吸收强度取决于两个因素，一是浓度或厚度，二是偶极矩的大小或吸光系数。如众所周知的比尔定律：吸光度（$A$）= $\varepsilon CL$，式中“$\varepsilon$”就是吸光系数，“$C$”可相当于我们涂片法中的某基团（或某辐射能量跃迁体系）的含量多少，“$L$”相当于样品厚度。十分明显，不同的基团或能量跃迁体系具有不同的偶极矩,具有不同的吸光系数。这个问题是光谱学上一个最基本的很重要的问题,也是分析化学尤其是定量分析的基本理论基础。只有在同一种纯化合物的对比分析中(例如不同浓度的定量分析),在一定浓度范围内,吸光系数才相同,那么,这种情况下才是“吸收强度取决于浓度”。因此,绝不能把吸收强度看成仅仅取决于一个变化因素。图 1 中蒽和菲的光谱对比很能说明这个问题:蒽与菲同分异构,同一分子式,$C_{14}H_{10}$,同是三环稠环芳烃,但是两张光谱图差别很大,1400 ~ 1600 $cm^{-1}$差别尤其大。固然,1600 $cm^{-1}$都是起源于芳环的伸缩振动,但是,菲的 1600 $cm^{-1}$吸收带的吸光度为蒽的吸光度 2 倍以上,这是因为菲的结构破坏了苯环的对称性,增强了分子的偶极矩,从而增大了吸光系数。因此,我们在使用 1600 $cm^{-1}$吸收带时有以下三点值得注意:

(1) 不能认为 1600 $cm^{-1}$的吸收越强就说明样品中芳烃含量越高,或者盲目地把它的强度作为芳烃含量高低的准则。只能说在一定条件下有可能含量较高。

(2) 不能认为缩合程度愈高,1600 $cm^{-1}$吸收就会愈强。实际上有可能与此恰恰相反。

(3) 不能认为 1460 $cm^{-1}$/1600 $cm^{-1}$大于或小于 1 就说明“稠环芳烃中侧链上烷链结构与芳核结构”究竟谁占优势。

因此,我们必须从多方面分析 1600 $cm^{-1}$的吸收强度,正确处理之。

对于 1460 $cm^{-1}$,除了上述问题外,还有一个问题,即芳环的干扰问题。芳环的伸缩振动可以反映在 1480 $cm^{-1}$,有时甚至到 1450 $cm^{-1}$。当 1600 $cm^{-1}$吸收强度较大时,这种干扰就会更严重。因为这个区域的吸收强度要大于 1600 $cm^{-1}$,图 1 所示光谱亦可说明。当然,对不含芳核结构的纯的饱和烃来说,并不存在该问题。

### 2.2 关于 ~1700 $cm^{-1}$

这个区域主要问题有两个,一是污染物的干扰十分严重。当然,如果存在污染的话,并不仅仅对 ~1700 $cm^{-1}$有干扰,对其他区域以至整个图谱也会有干扰。其所以在这个区域,把这个问题提出来,主要是污染物在这个区域吸光度可能很大,干扰就会更为突出。再说,其他区域,例如 1300 ~ 1000 $cm^{-1}$,我们应用就比较少;而对于另一些吸收带,污染物的吸光系数较小,影响的程度也就较小。

污染物主要是芳香脂类化合物,如邻苯二甲酸二丁酯。来源是多方面的,主要有试剂(如经常大量使用的氯仿)、滤纸以及塑料样品袋等。这类物质的污染将引起 ~1700 $cm^{-1}$(邻苯二甲酸二丁酯为 1740 $cm^{-1}$)的严重干扰。

另外一个问题是诸如芳烃样品的处理和长时间的放置，长时间地与空气接触，受空气中氧的氧化也是相当严重的，从而显著地增强 1700 $cm^{-1}$的吸收强度。

### 2.3 吸收强度比值的其他影响因素

根据试验（详见《方法试验小结》，国家地质总局中心实验室，1978.4）对吸收强度比值的影响，除了上述有关问题外，概括起来，主要的影响因素有：

（1）厚度的影响。涂片法制备样品，目视其样品厚度似乎差别不大，但是相对比较来说，不同涂片的相对厚度差异是不小的，无论何种吸收峰的比值，都受这种厚度差别的影响。只是影响的程度不一。

（2）计算方法的影响。上述方法试验小结对比了各种计算方法，可以看到相互之间的差别也是显著的。

（3）操作方法的影响。操作方法包括仪器操作方法与仪器分析条件，如狭缝程序的选取、100% 调节器的使用方法、参臂衰减等。

（4）仪器本身性能的影响。

## 3 关于红外光谱指标（或比值）选取的原则

红外光谱在石油地质中的应用虽然很广泛，国内外研究历史很长，在研究生油岩上亦如此，但是，指标成熟者不多。根据上述讨论，综合起来，应当遵循下列几条原则：

（1）地质意义和研究目的明确；

（2）化学的、地球化学的机制正确；

（3）光谱分析方法、样品处理分析方法准确。

本文原载于《石油实验地质》1978 年第 1 期

# 20. 中国“加瓦尔胚胎”与有机成油理论铁证生油剖面　一种新命名的岩石——“油泥灰岩”的命名申述

# China's “Garhwal embryo” and the organic generation theory irrefutable evidence of oil generation cross-section; A new naming rock— “Olefiant Marlstone Rock”, naming representations

何志高

本文在同一种理论基础上讨论以下几个有着紧密内在联系的问题:

## 1　中国“加瓦尔胚胎”

众所周知,加瓦尔(Ghawar)油田是世界上最大的油田,位于中东沙特阿拉伯,它的可采储量达107.4亿吨(http://bbs.zidonghua.com.cn),这个储量比美国(33.67亿吨)和中国(32.74亿吨)现今探明的储量加起来还要多;这个储量令很多人不可思议,甚至有人以此不可思议的巨量(称为“油海”)而从根本上否定有机成油理论,他们认为这是有机成油理论的最大弱点,有机成油理论解释不了这种油海的形成。由此可见,无论从极高储量的实际价值上或者从理论意义上而言,认识加瓦尔油田,以加瓦尔油田作为最优化的参照体,对比思考“特提斯(Tethys)”以及石油成因问题和中国的石油问题,是具有重要而又更具有特殊意义之举。

笔者于20世纪七八十年代在研究碳酸盐岩生油时发现了一个碳酸盐岩区的原生富油带,当时称为“窄而厚的富油相变过渡带”。30多年后的今天,笔者将这个碳酸盐岩区典型的相变过渡带中的特别富油的层段称为“加瓦尔胚胎”。

“胚胎”是生命科学中的概念,“加瓦尔胚胎”的寓意是具备可能发育成类似于加瓦尔油田的最初单元的物质基础,这种单元——胚胎及其系列,只要具备后续发展条件,就可以成为一个大型油气田。这种寓意性的对比,有利于对碳酸盐岩油气地质特殊性的认识;通过着重于石油地球化学的思考对比,还可能把“特提斯域富集油气”以及“世界性新月形油气富集带”等传统理念引导到一种新的理论领域,将传统的这些基于统计与描述为基础的理念上升为理论的、科学理性的认识,揭示加瓦尔这个世界最大油田神奇的奥秘;有利于最终解决石油成因问题。

30 多年前的 1978 年,笔者奉命参加地矿部的湘中石油地质攻关会战,在关士聪院士的直接指导下,担任地化组组长,开始系统研究碳酸盐岩的石油地球化学与生油问题,系统采集几条地化剖面的各类样品及其分析测试。与此同时,赴贵州八普,在刘敬谦先生的帮助下,得到了十分有价值的样品和资料;还采集了其他地区,包括中国北方碳酸盐岩,例如北京门头沟、甘肃南部的石炭系泥灰岩等大量而类型广泛的样品和资料,分析测试技术及其结果是严谨而可靠的。

笔者研究确立的中国“加瓦尔胚胎”是位于中国贵州省羊昌河地区羊 4 井井深从 851 ~ 865 m 地层的碳酸盐岩。样品分析结果如下:

酸解气态烃含量为 54 mL/kg 岩石,湿度为 25%,有机碳含量为0.35%,沥青含量为 0.076%,烃含量为 0.05%。

根据报道资料计算,加瓦尔油田每平方千米储油量为 $3\times10^6$ 吨(以 $A$ 表示);根据相关资料计算,贵州羊昌河地区富油相变过渡带(即中国“加瓦尔胚胎”系列)每平方千米储油量为 $2.54\times10^6$ 吨(以 $B$ 表示);“加瓦尔胚胎”(即羊 4 井井深 851 ~ 865 m 段)的储油量是上述 $A$ 值计算中所用实际含油量参数的 1.9 倍。

由上述 $A$、$B$ 数值可见,它们之间具有相近的可对比性。当然,二者均有正负误差偏差与不确定性。就 $B$ 值而言,主要是含油层厚度的确定,在该数的计算中,采取了偏低的估计值的计算,即油层厚度取 1000 m,而这套地层的实际厚度依据地表区域资料厚度超过 2000 m;其次,在含油量参数的计算中,对实验室分析资料族组成恢复到原油族组成,当然也会有偏差,笔者力求取最高的准确度,以缩小偏差。

羊昌河富油相变过渡带与加瓦尔油田产油层岩性相同,地质年代相近,均为中生代,前者为三叠纪,后者为侏罗纪;它们的有机碳含量也相近,羊 4 井近 900 m 的地层,比较均匀的采样点分布共 31 个样品,有机碳含量的算术平均值为 0.35%,而中东地区碳酸盐岩油源岩有机碳平均含量为 0.24%。

笔者研究对比后认为,羊昌河富油相变过渡带就单位面积而言,其富油性不会亚于加瓦尔油田的整体平均值，而羊 4 井 851 ~ 865 m 的富油碳酸盐岩，当之无愧地可以称之为“加瓦尔胚胎”。

## 2 有机生油理论铁证生油剖面

笔者赞成大师名言：“科学重视实证，玄学重视推理。”本文就是在过去已经发表的文章中报道过的“实证”的基础上，再拿出一种“实证”来，以丰富证明石油有机成因理论，而且认为这是一种“铁证”。

10 年前，笔者在《关于石油成因理论发展历史若干问题的思考与评论》一文中指出，有机成油理论证据数量很多，范围很广，研究程度很深，资料非常丰富；在这些证据中排在首位的是石油储量的地史分布与煤储量和油页岩储量的一致，铁证如山。可是，近年来，又有一些文章认为这还不是直接的证据，甚至连笔者与大庆油田微生物组合作的微生物直接生化反应形成石油烃类的实验证据也还不够。在微生物生化反应 96 h 的反应产物中，不仅气体湿度很高，正异构丁烷比值跟原油油溶气比值完全一致，而且检出 $C_6^+$ 的烃类含量达到1.31%。若所有这些证据还不够，笔者再向世界推出一条石油成因铁证剖面（表 1）。

表 1 有机成油理论铁证生油剖面主要资料汇集表

| 井深 Depth (m) | 岩性 Lithic | ΣHC (μL/kg rock) | 湿度 Humidity | Corg (%) | 沥青 A (%) | 烃 Hydrocarbon (%) | 油气显示 Display of oil and gas |
|---|---|---|---|---|---|---|---|
| 175 | 泥质灰岩 | 141506 | 30 | 0.38 | 0.042 | 0.027 | |
| 180 | 泥质灰岩 | 164804 | 32 | 0.32 | | | 187 ~ 205(m) 油苗甚多 |
| 215 | 灰质泥岩 | 8199 | 21 | 0.42 | 0.047 | 0.026 | |
| 237 | 砾状灰岩 | 41228 | 17 | 0.37 | 0.0079 | 0.006 | |
| 265 | 泥质灰岩 | 104252 | 33 | 0.34 | | | |
| 290 | 灰岩 | 54951 | 17 | 0.27 | 0.006 | 0.005 | |
| 308 | 泥质灰岩 | 144331 | 32 | 0.27 | | | |
| 321 | 泥灰岩 | 18348 | 21 | 0.20 | 0.047 | 0.033 | 265 ~ 275(m) 油苗特多 |
| 356 | 泥灰岩 | 12213 | 30 | 0.22 | 0.055 | 0.029 | |
| 366 | 泥灰岩 | 12081 | 20 | 0.28 | 0.032 | 0.019 | |
| 494 | 泥灰岩 | 54556 | 20 | 0.30 | 0.058 | 0.043 | |
| 510 | 灰质泥岩 | 23422 | 15.2 | 0.38 | 0.042 | 0.029 | |

续表

| 井深 Depth (m) | 岩性 Lithic | ΣHC (μL/kg rock) | 湿度 Humidity | Corg (%) | 沥青 A (%) | 烃 Hydrocarbon (%) | 油气显示 Display of oil and gas |
|---|---|---|---|---|---|---|---|
| 562 | 泥质灰岩 | 77401 | 28.3 | 0.31 | | | |
| 579 | 泥质灰岩 | 107683 | 28.4 | 0.43 | 0.024 | 0.021 | |
| 602 | 泥质灰岩 | 68194 | 19.0 | 0.35 | 0.047 | 0.037 | |
| 623 | 泥灰岩 | 40548 | 15.2 | 0.33 | | | |
| 640 | 泥质灰岩 | 80981 | 15.3 | 0.37 | 0.040 | 0.032 | |
| 661 | 泥质灰岩 | 26923 | 16.3 | 0.32 | | | 528 ~ 538(m) 油苗特多 |
| 688 | 泥质灰岩 | 41244 | 22.1 | 0.25 | 0.036 | 0.028 | |
| 715 | 灰质泥岩 | 75999 | 16.6 | 0.25 | 0.056 | 0.029 | |
| 749 | 泥灰岩 | 86395 | 21.7 | 0.37 | | | |
| 773 | 泥灰岩 | 80989 | 21.7 | 0.36 | | | |
| 791 | 泥灰岩 | 91420 | 15 | 0.46 | 0.045 | 0.032 | |
| 842 | 泥灰岩 | 23464 | 21.1 | 0.36 | | | |
| 850 | 泥灰岩 | 54195 | 25 | 0.35 | 0.076 | 0.050 | |
| 881 | 灰质泥岩 | 30820 | 16.2 | 0.39 | 0.034 | 0.025 | |
| 902 | 泥质灰岩 | 33668 | 17.2 | 0.48 | | | |
| 924 | 泥质灰岩 | 59491 | 13.7 | 0.53 | 0.048 | 0.039 | 851 ~ 865(m) 油苗甚多 |
| 951 | 灰质泥岩 | 8811 | 13.9 | 0.40 | | | |
| 966 | 泥灰岩 | 6450 | 11.9 | 0.36 | | | |
| 989 | 灰质泥岩 | 189754 | 27 | 0.43 | 0.011 | 0.006 | |

这条剖面就是贵州羊昌河地区羊 4 井从井深 175 ~ 989 m，集钻井录井、地质、地球化学相结合的综合剖面。这条剖面代表一个区域、地层连续累计 800 多米厚度、储集有数以千万吨的石油储量，其中的每一吨石油自生自储，是原生的。原油分析资料参见笔者相关文章。

证明它的原生性并不难，就是由无机酸分解碳酸盐矿物后释放出来的大量烃类，包括酸解气态烃、沥青 C，这些烃类就是原生的，这是一个在原理上最容易理解和最能胜于雄辩的事实。

实验分析过程是这样的：将样品混碎后，用有机溶剂将游离的、存在于孔洞缝隙中以及矿物颗粒之间的石油类物质全部抽洗完后，再加无机酸，此时矿物被分解，再用同一种有机溶剂、同一种条件下抽提。出来的石油类物质就是一类与碳酸盐矿物同时形成的被裹挟在矿物结晶体中的烃类，这就是“酸解烃类”的获得过程与确定的赋存状态。

众所周知,在常压条件下,碳酸钙的分解温度是约 800 ℃,碳酸镁、碳酸铁等的分解温度略低一些。这就是说,没有很高的温度,碳酸盐矿物不会分解,外界的非原生性的石油烃类是不可能出入于碳酸盐矿物内部的。绝对不是“天外来客”和“地下尖兵”,什么宇宙成因学说、费托合成、干酪根理论等统统会在这条铁证剖面事实面前得到检验。

剖面资料是 20 年前的,但尚未在正式出版物上发表,敬请各位同人前往贵州实地察看,取样进行研究。

## 3 一类新命名的岩石——油泥灰岩

从上述可见,羊 4 井揭示的有机成因理论铁证生油剖面的岩石是一类特殊的岩石,是真正的碳酸盐岩生油岩和储集岩生储成一体的岩石类,笔者给它命名为“油泥灰岩”,英文建议译为“Olefiant Marlstone Rock(简称 OMR)”。

这类岩石世界上没有人认识它、重视它,没有给它命名,没有给它应有的地位。这类岩石不同于油页岩,也不同于油砂,在石油地质学中没有给予碳酸盐岩生油岩一个名正言顺的名称。而实际上,碳酸盐岩给予人类的石油量是其他岩石不能比的,其贡献最大。前不久,阿拉伯-美国石油公司高级地质顾问 Abdulkader Afifiji 对加瓦尔油田的石油地质特征作了较为详细的介绍,他认为加瓦尔油田储油层是侏罗系“碎屑石灰岩”,生油岩是侏罗系富含有机质的“灰泥岩”。由此可见,加瓦尔这个世界上最大的油田,碳酸盐岩油田,权威学者还是应用传统的石油地质理论,误把主力生油岩当成储油岩,把非主力生油岩当成唯一的生油岩,淹没了油泥灰岩的功绩。

笔者在仔细研究了羊 4 井的全部资料后认为,碳酸盐岩油田主要的生油岩是一类生油的同时也是储油的泥灰岩,而不是“富含有机质的灰泥岩”,这种判断既适合羊 4 井,也适合加瓦尔油田和所有有一定规模储量的碳酸盐岩油田。从羊 4 井近 900 m 厚的 31 块样品的对比统计研究中清楚地可以看出,6 块泥岩类样品,除去一个较特殊的外,酸解气态烃含量和湿度平均值明显低于其他泥灰岩类,虽然有机碳含量确实是泥岩类高于灰岩类。笔者强调指出,成岩后的现今分析测试出来的有机碳实际上是地史时期沉积和成岩过程生化反应后剩余的有机物质,它含量高只能说明生化反应的成油过程进行得不彻底,生成油气的数量并不多。因此可以说,把“富含有机质的岩石”判定为生油岩的传统石油地质理论是一个错误观念与理论,当然也是 Abdulkader Afifiji 先生的一个错误。

判定识别油泥灰岩的最简易快速的方法和指标是酸解气态烃的湿度接

近高压地层原生油气藏的油溶气的湿度特征，笔者在《应用酸解气态烃评价碳酸盐生油岩》等文章中有详细论述。油泥灰岩的优劣与其他的岩石类一样也有级别等级之分。兼备生油岩与储油岩性能的油泥灰岩的综合评价当然需要多种综合方法和指标，这方面的研究无疑是石油地质学长期的任务。

## 4 观念与理论

其实，石油成因机制并不复杂高深。

“轻舟已过万重山！”，历经整整半个世纪毕生研究石油成因理论的笔者，今天想引用中国古代诗人这一名句来讲述一些容易被人接受的观念。

首先，笔者想特别强调，水在石油生成和油田形成中的极其重要、极其特别、极其巨大的作用。无水就无油。石油是在饱含水的地层、封闭的地质环境下生成、运移、聚集成藏的。凡正常原生油藏、大型油田以及世界上最大的加瓦尔油田都是如此。水的作用，微生物及酶所进行的生物化学反应是一个最关键的问题。石油成因问题和世界上许多其他科学问题一样，在没有解决之前是复杂的难题，而一旦突破之后，显现出来的问题并不复杂，也不深奥。

笔者认为，碳酸盐岩区石油的生成和油气藏的形成最能反映出水的作用，所以，笔者即以此为重点。水的作用有三：

（1）为沉积物中的微生物进行消化沉积有机物质提供水的介质环境，微生物自身也得以生存和繁衍。

（2）水与生物反应形成的二氧化碳一起与沉积物中的各种碳酸盐反应形成可溶于水的重碳酸盐。这个作用可谓神奇而重要。看似十分普通的化学反应，它一方面将碳酸盐碎屑、碳酸盐胶结物等溶解，为生成的油气创造储集空间和运移通道。例如，我们能亲眼看到的喀斯特地质现象和地质作用，地质景观。类似的大大小小的溶洞就是造就碳酸盐岩区“油海”的地下喀斯特。另一方面，使水介质的 pH 保持在一个比较高的适应于油气生成的数值，从而将全部的有机物质彻底转化为石油。例如，在沼气的生产中要保持一定的 pH 值一样。笔者提供的微生物在大庆油田所做的生油生化反应实验完全证实了这种控制酸度效应。

（3）油在水面上具有极易、极强、极快的运移聚集能力。在水面上，油气运移通道可以很狭窄，所需运移的压力差可以很小。在水界面上的石油运移真可谓不费吹灰之力，日行千米不为过。

由上述可见，石油不仅仅在水里生成，而且，油藏也是在水环境中运移聚集而成的。20 世纪 90 年代初，美国学者 Peters 等应用 $^{14}C$ 测年法，确定加利福尼亚海湾 Guaymas 盆地的石油和碳酸盐的生成年龄为 4240 ~ 5705 年。这是非常有意义的重大的研究成果与发现，笔者完全相信 Peters 等的结论的可靠性。同时，笔者还相信，4240 ~ 5705 年可能还不是最短时间的世界纪录。石油从生成到聚集成藏，对条件合适的油气藏而言，可以是一个非常短的时期，笔者在过去研究中也发现并在文章中报道过，例如《雏形油气圈闭》等。而且，笔者还认为，更具有特殊意义的是 Peters 等研究的石油的生成年龄与碳酸盐的生成年龄是同样的，也就是说它们是同时生成的，与笔者前述理论完全一致。

不久前，笔者看到有一种观点，即 $C_{10}^{+}$ 烃类未能在实验室内获得，从而否定有机成因理论。笔者认为，在实验室 10 个碳原子的烷烃类制取早已不是什么难题；理论上的研究，脂肪酸的脱羧反应的热力学分析以及实验条件研究都较充分。

在此也顺便提及一个有关的问题：一些对煤层产状，尤其是那些煤层薄到被称为“煤绒”的产状，用有机成因理论解释相“矛盾”的质疑者也是因为没有看到煤矿中的“树根”就认为煤不是由高等植物形成的，如此等等。笔者认为，这些质疑者也没有看到水和微生物对成煤有机物质改造的巨大作用。

本文原载于《石油知识》总 148 期(2011)

# 21. 一种提高原油采收率的新方法——马弗阿法

## A new method for enhanced oil recovery—MFA method

何志高

(地质矿产部石油地质中心实验室,江苏无锡,214151)

**摘要:** 该文报道一种提高原油采收率的新方法——马弗阿(MFA)法的方法原理、理论基础、微生物模拟实验及提高原油采收率的模拟实验、增产效果估算等资料。同时还探讨了该方法实施后在地层中的生气量与原油量的关系等问题。

**关键词:** 原油采收率;马弗阿法;微生物;专用流体

历来,提高原油采收率为很多国家所重视,因为提高采收率与找到新的油气田具有同样的意义。20 世纪 80 年代中期,美国已有 750 个提高原油收率的方案,反映了人们在寻求新的、有效的和最佳经济效益的提高原油收率方法上所做出的巨大而艰难的努力,也表明了探索新方法的重要性。在原苏联,现阶段发展石油工业的特点之一就是增加开发晚期油藏的采油量。

马弗阿(MFA)法是一种新方法(已申请中国专利),它可以归属为生物采油范畴,但与一般的生物采油法有着重要的区别。该法吸取了现有各种方法机制上的精髓,以最新石油成因理论、油田水的研究资料以及沼气发酵实践、微生物模拟实验为基础,并进行了有关的提高原油采收率的实验室模拟实验,预计可提高二次采油率约 5% 。该法具有永久性效益,是长期对提高采收率起作用的因素。该法与现有诸方法相比,有一个重要的差异,即能够增加新生的天然气产量;另一个重要特点是,可以在同一油井(含注水井)、同一层位多次实施,对其他增产措施无干扰,对油层无污染,没有一般微生物方法的弊端;再者,该法实施技术不复杂、成本低。显然,该法的优势是突出的。

本法主要适应于油层温度不超过 100 ℃的原生油藏的二(三)次采油。

## 1 方法及其原理简述

配制一种特殊的前置液和一种发酵液与脂肪酸的混合物,随同压裂用砂液(实际上是按先后次序)注入油层,或将前置液和混合物分别压入注水井对应地层(注入水应调节 pH),经过一段时间的反应,直接或间接产生油溶气,增加原油的流动性,即可按正常采油程序开放油层采油。

实施例:一口压裂工程油井,可取沼气发酵液,加入乙酸,pH 调整至小于 6,其配方用料为:沼气发酵液 $>1\ m^3$,乙酸 $1\sim10$ kg,前置液 $>1\ m^3$。

## 2 基于油气成因新理论

新的油气成因理论湿封闭体系综合效应成油论[1-3]指出,工业规模的石油并非由赋存于“生油岩”中的分散有机质——干酪根热降解形成的,其形成的时间、地点、条件、方式可简述如下:

有机物质(包括陆源的与水生的)随同各类碎屑物质沉积后,在失去游离水前,受到生物化学作用以及物理分异作用的影响,经过改造,重新分配,其结果是最轻的、分散细粒状的以及胶状的有机物聚集在细粒沉积物(如泥质)的表面,并和上部继续沉积覆盖的粗碎屑(如砂)相混,从而处于含水的多孔介质之中。这种孔隙中的有机物(包括粗碎屑沉积物中的有机质)在继续进行微生物的“发酵”作用的时候,沉积覆盖另一个沉积旋回的细粒沉积物,此种沉积物作为盖层,从而形成一个隔氧的饱含水的封闭体系。在这个湿封闭体系中,有机物经微生物的长期彻底改造,形成各种脂肪酸、类石油胶质、类石油沥青质以及甲烷气体等。这个时期可以称为生油的准备阶段。

在湿封闭体系中,继形成脂肪酸的生化反应之后,开始了脂肪酸脱羧的生物化学阶段。在此阶段,形成各种轻重气态烃以及高分子烃类。这些烃类,加上胶质与沥青质,构成早期生成的石油。

随着沉积物埋深作用增强,温度相应增高,类石油胶质、类石油沥青质以及早期形成的石油胶质、石油沥青质热降解形成一部分烃类。这些烃类与生物作用形成的烃类构成了全部石油烃类的总和。温度相应增高的结果,除了这种增加烃类的效应外,还有增强压力的效应和降低石油黏度、增加石油流动性的效应。温度增高还将导致烃类汽化程度增加,气体膨胀,再导致增大封闭系统的内压力。由于压力增加,反过来又有气态烃类溶解于液相油中的量增加,从而使石油运移能力增强。

这种成油的体系对外界环境来说是封闭的，对体系内部来说却是高渗透性的。在这种湿封闭体系中形成的石油气态烃（伴生气）、液态烃以及胶质、沥青质在温度与压力综合效应的影响下，石油各组分分子间内聚力增强，相互溶解性增强，黏度降低，运移能力增强，分散状的石油聚集成为一个整体，并受岩性等条件的影响，最终形成规模不等的油（气）藏。

由此成油理论可知，在成油的准备阶段，有机质形成脂肪酸类。在早期成油作用中，脂肪酸脱羧形成烃类的同时，形成二氧化碳。由于整个成油过程均是在含水的多孔介质之中进行的，二氧化碳必定与砂岩中的碳酸盐胶结物（或碳酸盐类矿物）反应，形成高 pH 的碱性水。水的高碱性导致微生物进一步脱羧作用减弱以至完全停止。因此，原生油藏必然残留着一定数量的未反应的脂肪酸。

原生油藏的油田水中富含脂肪酸以及油田水中普遍含脂肪酸较高的事实在 20 世纪 60 年代后即有分析测试与地化研究资料报道。根据中国大庆油田油田水的资料，脂肪酸含量可高达 2 g/L 以上。

## 3 基于沼气发酵理论与实践

沼气发酵（marsh gas fermentation）是在一个厌氧的系统中，各种不同类型的微生物相互协同、相互制约完成的一种极其复杂的生物化学过程，以实现各种微生物的代谢[4,5]。

### 3.1 沼气发酵的热力学

沼气发酵的阶段理论有二阶段论、三阶段论、四阶段论三种理论。二阶段论者认为，第一阶段产酸，第二阶段产甲烷；三阶段论者认为，第一阶段发酵菌使碳水化合物水解，第二阶段产氢、产醋酸，第三阶段产甲烷；四阶段论是将每个阶段独特的微生物菌群作为划分阶段的根据。研究这些阶段各自的热力学似乎没有必要，我们所需要的主要是各种生成甲烷反应的热力学性质。

沼气发酵产物主要是甲烷和二氧化碳，以乙酸代表反应物（或代表产酸菌分解代谢产物），其化学反应式如下：

$$CH_3COOH(液) \longrightarrow CH_4(气) + CO_2(气) \quad \cdots\cdots\cdots\cdots\cdots \quad (1)$$

根据计算，该反应式的 $\Delta G^0 = -52.7\ \mathrm{kJ \cdot mol^{-1}}$，$\Delta H = -18.66\ \mathrm{kJ \cdot mol^{-1}}$，$\Delta S = 240.0\ \mathrm{J \cdot deg^{-1} \cdot mol^{-1}}$。由此可见，该反应是放热、熵增、自由焓降低的反应，是在热力学标准状况下能自发进行的反应。其 $\Delta G$ 是在生物生理变化能量级（$33.5 \sim 50.2\ \mathrm{kJ \cdot mol^{-1}}$）范围。这就是说，整个发酵体系在能量上是符

合化学热力学的理论原则的。

根据 M. P. Bryant[6]的研究，在沼气发酵过程中，各主要的单个反应，其 $\Delta G^0 \leqslant 0$，如：

$4H_2 + HCO_3^- + H^+ = CH_4 + 3H_2O$ ……(2)，$\Delta G^0 = -1354$ kJ·反应$^{-1}$

$4HCOO^- + H_2O + H^+ = CH_4 + 3HCO_3^-$ ……(3)，$\Delta G^0 = -130.4$ kJ·反应$^{-1}$

$4CH_3OH = 3CH_4 + HCO_3^- + H^+ + H_2O$……(4)，$\Delta G^0 = -314.3$ kJ·反应$^{-1}$

$CH_3 \cdot CH_2 \cdot COO^- + H_2 = CH_3COO^- + CH_4$……(5)

$\Delta G^0 = -102.0$ kJ·反应$^{-1}$

### 3.2 沼气发酵的动力学

根据 Y. R. Chen 等[7]研究，发酵系统的反应速度由是细菌细胞量的变化速度，和基质浓度的变化速度所支配，即

$$dM/dt = \mu M - M/\theta \quad \cdots\cdots(6)$$

$$dS/dt = -F + (S_0 - S)/\theta \quad \cdots\cdots(7)$$

$M$ 是细胞物质量浓度(量/体积)，$\mu$ 是微生物的比生长率(时间$^{-1}$)，$F$ 是体积基质利用率(物质体积$^{-1}$·时间$^{-1}$)，$S$ 是生物降解的流出液基质量浓度(量/体积)，$S_0$ 是生物降解的流入准基质量浓度(量/体积)，$\theta$ 是发酵液滞留时间，$t$ 是时间。

这些动力学方程因机制与本方法(一次性加料)不完全相符，所以不适用于本方法，仅可作为参考。本文所研究的对象不是产甲烷，而是微生物的量。根据钱泽澍等[4]的资料，产甲烷菌数量在第一个 15～20 d 可翻一番，从 $10^{-2}$/mL 增加到 $10^8$/mL，发酵细菌数量可达 $5 \times 10^{10}$/mL。

### 3.3 微生物优势种群的演替

综合前人的研究，在沼气发酵的初期，尚未形成还原环境，由于水中、发酵物料中存在游离氧，又与空气接触，因此，喜氧菌较多，其生化作用仍然占有很重要的地位。在不断添加物料的情况下，喜氧菌及其生化作用只能逐渐减少、减弱，而不能得到根除；在一次性加料的发酵中、喜氧生化作用显然会较快地减弱，直至沼气生成量从顶峰下落，喜氧菌方可消耗殆尽。此时，厌氧菌及其生化作用占有绝对的优势。在厌氧发酵中，由于发酵原料不同，有水溶性的如糖类、氨基酸等，有不易水解的如纤维素、果胶、芳香族化合物等，还有难于水解的如木质素等，因而对于细菌可作为能源被利用的时间就有先后次序之分，这也是产气高峰出现双峰的原因。一般来说，温度越低，双峰愈不明显。

## 4 微生物模拟实验研究

微生物模拟实验方法采用氮气排水法,完全消除空气(或氧)的干扰,在氮气环境中进行,各类发酵物、营养物、盐、酸(碱)液均经针头注入。

### 4.1 脂肪酸发酵试验

表1是深层井水介质加入脂肪酸的发酵试验结果。从表1可见:

(1) 介质pH4~10条件下,$CH_4$生成基本一致;pH>10条件下,$CH_4$生成量显著减少甚至停止生成。

(2) 介质pH4~10条件下,二氧化碳均有生成;pH<4条件下,生成量显著减少;PH过高亦略有降低。

(3) pH>10条件下,有很多的$H_2$生成。

(4) 二氧化碳生成量远远高于$CH_4$。

该试验发酵时间(室温)长达3个月,表1中的产气绝对量均不及本专题研究(本文)其他模拟试验多,其原因是深层井水中微生物数量较少之故。

**表1 深层井水与混合脂肪酸发酵模拟试验结果**

| 编号 | 脂肪酸(mL) | 营养物(mL) | NaCl | pH | $CH_4$(%) | $H_2$(%) | $CO_2$(%) | 烯烃 |
|---|---|---|---|---|---|---|---|---|
| B1 | 不加 | 0.095 | 不加 | 6.5 | 0.026 | 未检出 | 20.6 | 未检出 |
| B2 | 0.1 | 0.095 | 不加 | 6.5 | 0.115 | 1.29 | 19.45 | 痕 |
| B3 | 0.1 | 0.095 | 1% 0.2 mL | 6.5 | 0.12 | 0.50 | 20.63 | 痕 |
| B4 | 0.1 | 0.095 | 1% 2 mL | 6.5 | 0.13 | 1.24 | 19.35 | 痕 |
| B5 | 0.1 | 0.095 | 1% 2 mL | 10 | 0.13 | 0.89 | 18.44 | 痕 |
| B6 | 0.1 | 0.095 | 1% 2 mL | 11 | 0.009 | 18.63 | 18.38 | 未检出 |
| B7 | 0.1 | 0.095 | 1% 2 mL | 4 | 0.10 | 未检出 | 14.17 | 痕 |

### 4.2 不同发酵液与不同脂肪酸的中-低温生化反应

表2列出了不同原料、不同条件与不同脂肪酸中-低温(35 ℃)生化反应结果。

表 2　中-低温(35 ℃)生化反应结果汇总

| 样品号码 | 母液体积(mL) | 脂肪酸类及其量(μL) | pH | $CH_4$(%) | $C_2$(%) | $CO_2$(%) | $H_2$(%) |
|---|---|---|---|---|---|---|---|
| C1-01 | 10 | 乙酸 30 | 6.5 | 0.25 | 未检出 | 1.62 | 未检出 |
| C2-01 | 10 | 乙酸 30 | 6.5 | 0.02 | 未检出 | 1.23 | 未检出 |
| C3-01 | 7.7 | 乙酸 30 | 6.5 | 0.32 | 未检出 | 1.32 | 未检出 |
| J-01 | 10 | 乙酸 30 | 6.5 | 0.05 | 未检出 | 0.83 | 未检出 |
| N1-01 | 5 | 乙酸 30 | 6.5 | 0.11 | 未检出 | 1.31 | 未检出 |
| C1-02 | 10 | 丙酸 10 | 6.5 | 0.26 | 未检出 | 1.20 | 未检出 |
| C2-02 | 10 | 丙酸 10 | 6.5 | 0.02 | 未检出 | 0.70 | 未检出 |
| C3-02 | 8 | 丙酸 10 | 6.5 | 0.99 | 未检出 | 1.32 | 未检出 |
| J-02 | 10 | 丙酸 10 | 6.5 | 0.20 | 未检出 | 1.55 | 未检出 |
| N-02 | 5 | 丙酸 10 | 6.5 | 0.05 | 未检出 | 0.87 | 未检出 |
| C1-03 | 10 | 正丁酸 10 | 6.5 | 0.13 | 未检出 | 2.45 | 未检出 |
| C2-03 | 10 | 正丁酸 10 | 6. 5 | 0.01 | 未检出 | 0.61 | 未检出 |
| C3-03 | 10 | 正丁酸 10 | 6.5 | 0.96 | 未检出 | 1.23 | 未检出 |
| J-03 | 6 | 正丁酸 10 | 6.5 | 0.16 | 未检出 | 1.19 | 未检出 |
| N1-03 | 5 | 正丁酸 10 | 6.5 | 0.04 | 未检出 | 1.69 | 未检出 |

样品:C1 为酒糟 35 ℃发酵液;C2 为酒糟 55 ℃发酵液;C3 为猪粪发酵液(反应中取样);J 为猪粪发酵液(反应后废液);N 为柠檬酸发酵废液

从表 2 可见:

(1)乙酸与各类发酵液生化反应生成甲烷量比丙、丁酸均少,生成二氧化碳的量以丙酸最少。5 种实验平均结果如下:

| | $CH_4$(体积分数,%) | $CO_2$(体积分数,%) | $CO_2/CH_4$ |
|---|---|---|---|
| 乙酸 | 0.15 | 1.30 | 8.7 |
| 丙酸 | 0.25 | 1.01 | 4.0 |
| 丁酸 | 0.26 | 1.43 | 5.5 |

(2)35 ℃酒糟发酵液及乙酸反应生成 $CH_4$ 量比 55 ℃发酵液高,即 35 ℃为 0.21%,55 ℃为 0.02%。

(3)以猪粪为原料(C3)的发酵液与各种脂肪酸反应,平均产 $CH_4$ 气量最高为 0.76%。其他分列如下:

| | $CH_4$（体积分数，%） | $CO_2$（体积分数，%） | $CO_2/CH_4$ |
|---|---|---|---|
| 35 ℃酒糟发酵液（C1）平均值 | 0.21 | 1.82 | 8.7 |
| 55 ℃酒糟发酵液（C2）平均值 | 0.02 | 0.85 | 42.5 |
| 柠檬酸发酵液（N）平均值 | 0.14 | 1.29 | 9.2 |
| 猪粪发酵后废液（J）平均值 | 0.07 | 1.19 | 17.0 |
| 猪粪发酵液（C3）平均值 | 0.76 | 1.29 | 1.7 |

由此可见，对产生 $CO_2$ 气体来说，35 ℃的酒糟发酵液高于 55 ℃发酵液的产气量。

### 4.3 不同发酵液与脂肪酸的中-高温生化反应

将上述中-低温反应模拟试验样品升高温度至 45 ℃（5 d），再进行分析，其结果如表 3 所示。

对比表 3 与表 4，由于温度升高 10 ℃，$CH_4$ 量无大变化，说明 35 ℃条件已完成该类生化反应。而升高温度后，$CO_2$ 量有明显降低，推测系二氧化碳在溶液中的溶解度增大，气态二氧化碳进入液相之中所致。此外，还可能有不同微生物最适宜发酵温度的差异性问题。

**表 3 中-高温（45 ℃）生化反应结果汇总**

| 样品号码 | 母液体积（mL） | 脂肪酸类及其量（μL） | pH | $CH_4$（%） | $C_2$（%） | $CO_2$（%） | $H_2$（%） |
|---|---|---|---|---|---|---|---|
| C1 - 11 | 10 | 乙酸 30 | 6.5 | 0.24 | 未检出 | 0.80 | 未检出 |
| C2 - 11 | 10 | 乙酸 30 | 6.5 | 0.02 | 未检出 | 0.51 | 未检出 |
| C3 - 11 | 7.7 | 乙酸 30 | 6.5 | 0.28 | 未检出 | 0.44 | 未检出 |
| J - 11 | 10 | 乙酸 30 | 6.5 | 未检出 | 未检出 | 0.85 | 未检出 |
| N1 - 11 | 5 | 乙酸 30 | 6.5 | 0.03 | 未检出 | 0.34 | 未检出 |
| C1 - 12 | 10 | 丙酸 10 | 6.5 | 0.27 | 未检出 | 0.55 | 未检出 |
| C2 - 12 | 10 | 丙酸 10 | 6.5 | 0.02 | 未检出 | 0.29 | 未检出 |
| C3 - 12 | 8 | 丙酸 10 | 6.5 | 1.02 | 未检出 | 0.29 | 未检出 |
| J - 12 | 10 | 丙酸 10 | 6.5 | 0.17 | 未检出 | 0.86 | 未检出 |
| N - 12 | 5 | 丙酸 10 | 6.5 | 0.05 | 未检出 | 0.38 | 未检出 |
| C1 - 13 | 10 | 正丁酸 10 | 6.5 | 0.14 | 未检出 | 1.12 | 未检出 |
| C2 - 13 | 10 | 正丁酸 10 | 6.5 | 0.01 | 未检出 | 0.27 | 未检出 |

续表

| 样品号码 | 母液体积(mL) | 脂肪酸类及其量(μL) | pH | $CH_4$(%) | $C_2$(%) | $CO_2$(%) | $H_2$(%) |
|---|---|---|---|---|---|---|---|
| C3-13 | 10 | 正丁酸 10 | 6.5 | 0.01 | 未检出 | 0.47 | 未检出 |
| J-13 | 6 | 正丁酸 10 | 6.5 | 0.17 | 未检出 | 0.57 | 未检出 |
| N1-13 | 5 | 正丁酸 10 | 6.5 | 0.03 | 未检出 | 0.82 | 未检出 |

**表 4 同一样品低温至高温在不同条件下生化反应结果**

| 试验号 | 发酵液量(mL) | 加入脂肪酸种类及其量 | 温度(℃) | 时间(h) | pH | $CH_4$(%) | $C_2$(%) | $CO_2$(%) | $H_2$(%) | 备 注 |
|---|---|---|---|---|---|---|---|---|---|---|
| 0628 | 7.7 | 乙酸 30 μL | 35 | 50 | ≈6.5 | 0.32 | 未检出 | 1.32 | 未检出 | 样品:成都沼气研究所猪粪发酵液。混合酸:乙酸1000 μL,丙酸200 μL,正丁酸100 μL。 |
| 0633 | 8.0 | 丙酸 10 μL | 35 | 50 | ≈6.5 | 0.99 | 未检出 | 1.32 | 未检出 | |
| 0639 | 8.0 | 正丁酸 10 μL | 35 | 50 | ≈6.5 | 0.96 | 未检出 | 1.23 | 未检出 | |
| 0644 | 7.7 | 乙酸 30 μL | 45 | 50 | ≈6.5 | 0.28 | 未检出 | 0.44 | 未检出 | |
| 0649 | 8.0 | 丙酸 10 μL | 45 | 50 | ≈6.5 | 1.02 | 未检出 | 0.29 | 未检出 | |
| 0654 | 8.0 | 正丁酸 10 μL | 45 | 50 | ≈6.5 | 0.91 | 未检出 | 0.47 | 未检出 | |
| 0659 | 50 | 混合酸 0.2 mL | 45 | 96 | 6 | 0.35 | 未检出 | 33.5 | 0.48 | |
| 0662 | 50 | 乙酸 0.4 mL | 45 | 96 | 6 | 0.30 | 未检出 | 26.4 | 0.42 | |
| 0663 | 50 | 丙酸 0.2 mL | 45 | 96 | 6 | 0.13 | 未检出 | 27.6 | 0.23 | |
| 0664 | 50 | 正丁酸 0.2 mL | 45 | 96 | 6 | 0.11 | 未检出 | 25.3 | 0.32 | |
| 0665 | 50 | 混合酸 0.2 mL | 60 | 96 | 8.5 | 0.27 | 未检出 | 未检出 | 0.38 | |
| 0666 | 50 | 乙酸 0.4 mL | 60 | 96 | 9 | 0.35 | 未检出 | 未检出 | 0.89 | |
| 664-1 | 50 | 丙酸 0.2 mL | 60 | 96 | 9.5 | 0.18 | 未检出 | 未检出 | 0.47 | |
| 664-2 | 50 | 正丁酸 0.2 mL | 60 | 96 | 9.5 | 0.15 | 未检出 | 未检出 | 0.75 | |

## 5 提高原油采收率的模拟实验及其效果

本实验的目的是研究气体进入原油油相后,导致原油黏度的变化程度,以证实提高采收率的效果。

### 5.1 原油黏度变化实验

样品:陕甘宁原油(三叠系)。

实验条件:25 ℃,$1\times10^5$ Pa(约 1 个大气压)。

结果如表 5 所示。

表 5　原油黏度变化试验及其结果

| 原油流出时间(S)　测定<br>实验样品 | 1 | 2 | 3 | 4 | 平均 |
|---|---|---|---|---|---|
| 脱水原油 | 34.5 | 35.0 | 34.7 | 34.2 | 34.6 |
| 原油/气体 12∶1（体积比） | 34.0 | 33.8 | 33.8 | | 33.9 |

由表 5 可见，加入气体的原油比未加气体的原油从试验仪器中流出时间减少，表明原油黏度降低。

### 5.2　增加原油采收率模拟试验

试验方法：采收率增加量是通过让原油及加入气体后的油样流经特定的模拟装置，记录一定时间从多孔介质中继续流出的原油量。其结果如下：

| | |
|---|---|
| 未加入气体的原油残留量(mL) | 0.501 |
| 加入气体的原油残留量(mL) | 0.474 |
| 残留减少量(mL) | 0.027 |
| 提高原油收率 | 5.4% |

从上述数据可见，加入气体的油样残留量比不加入气体的原油残留量显著减少，提高原油收率的效果是显著的。

其实，以往的石油地质勘探和油田开发的大量试验和生产实践早已表明：

(1) 地下原油与地面（已分泌出油溶气的）原油，在相同温度条件下，地下原油因含油溶气，其黏度较小。

(2) 人工从地面向地层内注入气体（包括二氧化碳和各种烃类气），原油采收率提高。

因此，这两个早已成定论的问题是无须重复试验与赘述的。

## 6　产气量的估算

### 6.1　沼气发酵中产气量的几种估算法

(1) 布什威尔-麦尔夫法[4]（Buswell and Mueller）

该法的化学反应式为：

$$C_nH_aO_b + (n - a/4 - b/2)H_2O = (n/2 - a/2 + b/2)CO_2 + (n/2 + a/8 - b/4)CH_4 \quad \cdots\cdots\cdots (8)$$

若以葡萄糖为反应物（$H_2O$ 作为介质），则可写成：

$$C_6H_{12}O_6 + 0H_2O = 3CO_2 + 3CH_4 \quad \cdots\cdots\cdots\cdots\cdots\cdots\cdots (9)$$

由(9)式可见,1 分子葡萄糖产生 3 分子甲烷,或者葡萄糖中 6 个原子碳产生 3 个原子甲烷碳,或 180 g 葡萄糖产生甲烷 67.2 L。

(2) SHG 法[8]

SHG(R. J. Smith, M. E. Hein and T. H. Greinet)法对沼气发酵产气量计算如下式:

$$C_6H_{12}O_6 = 3CH_4 + 3CO_2 \quad \cdots\cdots\cdots\cdots (10)$$

(3) HH 法[9]

HH(R. Horton and D. L. Hawkes)法对 $C_{16}$ 一元脂肪酸 $CH_3(CH_2)_{14}COOH$ 的产气量计算为每升产气 1.4 $m^3$,即:

$$(22.4 \times 16)/256 = 1.4\ m^3 \quad \cdots\cdots\cdots\cdots (11)$$

按这种计算,本文笔者归纳其定量反应式如下:

$$CH_3(CH_2)_{14}COOH + (n-1)O_2 = nCO_2 + pCH_4 (n+p=16) \cdots\cdots (12)$$

综上所述 3 种方法,糖类、脂肪酸等物质的产气量,若以碳原子数在各反应式中的变化作估计,其产气率($F$)可表述为

$$0.5 \leqslant F \leqslant 1 \quad \cdots\cdots\cdots\cdots (13)$$

### 6.2 本方法产气量的估算

参考上述沼气发酵产气量估算中 $F$ 值,令本方法产气物源的产气率或产气反应效率系数为 $f$。当 $f=1$ 时,表示全部碳源转化为气体(含甲烷和二氧化碳);当 $f=0.5$ 时,则仅有一半的碳源转化为气体。前已述及,本方法产气物源有 3 种,即加入的沼气发酵液、产气增效液及地层水中的可产气的脂肪酸等。这 3 种物源的产气量分别计算如下:

(1) 沼气发酵液,$Q_A$

根据研究,沼气发酵液中,脂肪酸量可达 2 g/L,加上醇-酯类等,换算到乙酸当量,其产气物总量可达 5 g/L。按乙酸全部转换为甲烷与二氧化碳气体(即 $f=1$)计算,每立方米液产体量约为 3.7 立方米气,所以,

$$Q_A \leqslant 3.7\ kf (m^3) \quad \cdots\cdots\cdots\cdots (13)$$

式(13)中,$K$ 为发酵液体积,单位为 $m^3$;$f$ 为反应效率系数,$0 < f \leqslant 1$。

(2) 产气增效液,$Q_B$

产气增效液,乙酸,其本身亦可作为产气源,按每次用量 10 kg 计算,其产气量

$$Q_B \leqslant 7.4f\ m^3 \quad \cdots\cdots\cdots\cdots (14)$$

(3) 地层水中的可产气物,$Q_C$

根据油田水中脂肪酸含量的研究资料,油田水最高可达 2.5 g/L 的脂肪酸,此外,还有其他可产气的有机物质。按中等含量 2 g/L 计算,每立方米油

田水产气量约为 1.48$f$。油田水产气量($Y$)是岩石孔隙度($P$)、水饱和度($W$)、油层体积($V$,单位 $m^3$)及反应效率($f$)的函数,即

$Y=\Phi(P,W,V,F)$ ………………………………………… (15)

设 $P=0.15$, $W=0.4$, $V=3.1\times10^4$, $Q_C=1.48\times0.15\times0.4\times3.1\times10^4=2.7\times10^3 f\ m^3$ ………………………………………… (16)

由上述计算可知,能够对油层中原油的溶解气量产生影响的因素主要是地层水中脂肪酸等产气物源。

## 7 产气量与原油数量的关系

与上述 $Y$ 函数计算中同样体积油层相对应的原油体积,若以油饱和度0.4计算,则原油体积($Y_0$)为

$Y_0=3.1\times10^4\times0.15\times0.4=1.8\times10^3\ m^3$ ………………………… (17)

若产气量 $Q_C$ 的计算中,反应效率 $f$ 取 0.5,则 $Q_C$ 为 $1.35\times10^3$。再加 $Q_A$ 与 $Q_B$,则总产气量的气体体积 $Q$ 将与油的体积 $Y_0$ 相近,即

$Q\approx Y_0$,或者 $Q/Y_0\approx1$ ………………………………………… (18)

这种油气比值关系将大大高于本文中模拟采油回收率试验中的油气比。从此种意义可以判断,实际的效果将会优于实验室模拟实验。当然,最终的效果检验尚需在油田中进行实验和实施。

## 8 结束语

在提高原油采收率方法的发展历史上,每一种方法均经历了一个较长时间的、反复的、克服着各种理论上和实际中遇到的问题的不断充实和完善的过程。马弗阿法亦将无例外地要遵循这种普遍的规律,走过漫长而曲折的道路。但笔者相信,它必将开拓一个新的领域,在采油的历史上闪烁出光辉的功绩。

### 参考文献

[1] He Zhigao. On the origin of petroleum hydrocarbon cases[J]. Acta Geologica Sinica, 1986,60(4):57-65.

[2] 何志高. 湿封闭体系综合效应成油论,石油实验地质,1988,10(4):315-324.

[3] He Zhigao. A model of oil-generation in a waterlogged and closed system[J]. J Southeast Asian Earth Sci, 1991,5(1-4):465-469.

[4] 钱泽澍,闵航. 沼气发酵微生物学[M]. 杭州:浙江科学技术出版社,1986.

[5] (日)植村定治郎,相田浩. 发酵与微生物[M]. 天津市工业微生物研究所,译. 北京:科学出版社,1979.

[6] Bryant MP,. Microbial methane prodution—thoretical aspects[J]. J Anim Sci, 1979, 48(1):193 - 201.

[7] Chen YR, Hashimoto AG. Kinetics of methane fermentation[J]. Biotechnol Bioeng Syrup, 1978(8):269 - 282.

[8] Smith RJ, Hein ME, Greiner TH Experimental methane production from animal excreta in pilot-scale and farm-size units[J]. J Anim Sci, 1979, 48(1):202 - 217.

[9] Horton R, Hawkes DL. Anaeroblic digester design fundamentals [J]. Process Biochemistry, 1979, 14(9):12 - 16.

# 22. A new method for enhanced oil recovery—MFA method

Patent of P. R. China: Appl. No. 8810537. 0

C. N. 1035702A

Auth. 1992. 02. 05

MFA, a new method for enhanced oil recovery, collects quintessence of various methods, bases on data of recent theory of oil generation, synthesizes the relevant data of marsh ferment and simulated test by microorganisms.

The method is to press a "special fluid" into oil-bearing stratum, which contains rich microorganisms which can produce methane and carbon dioxide gases, reacting with aliphatic acids, alcohols and greases and so on. The gases decrease specific gravity and viscosity of oil and increase fluidity of oil and decrease remnant oil, due to those gases dissolved (or mixed) in oil. Thereby it attains goal to increase production of oil.

The method is applied to second (or third) recovery of original oil deposit which the depth is less than 3000 m and the temperature less than 100 ℃.

The method may apply alone to oil well or injecting water well, and also at the earlier or later stage of other EOR methods.

The method is carried out uncomplicated and fairly economical. It has not any bad influence to other EOR methods and has not side effects.

The most important thing is effective. It may raise the gather productivity of second recovery about 5%. The outstanding characteristic is that the method can be used repeatedly.

发明名称:一种提高原油收率的新方法——马弗阿法

发明人:何志高

专利号(申请号):88105374.0

专利申请日:1980 年 3 月 12 日

专利权人:何志高

发明

第 13321 号

共 2 页

该发明已由本局依照中华人民共和国专利法进行审查,决定授予专利权

中华人民共和国专利局局长 高卢麟

1992 年 2 月 5 日

# 23. 原油组成一览表

List of crude oil composition

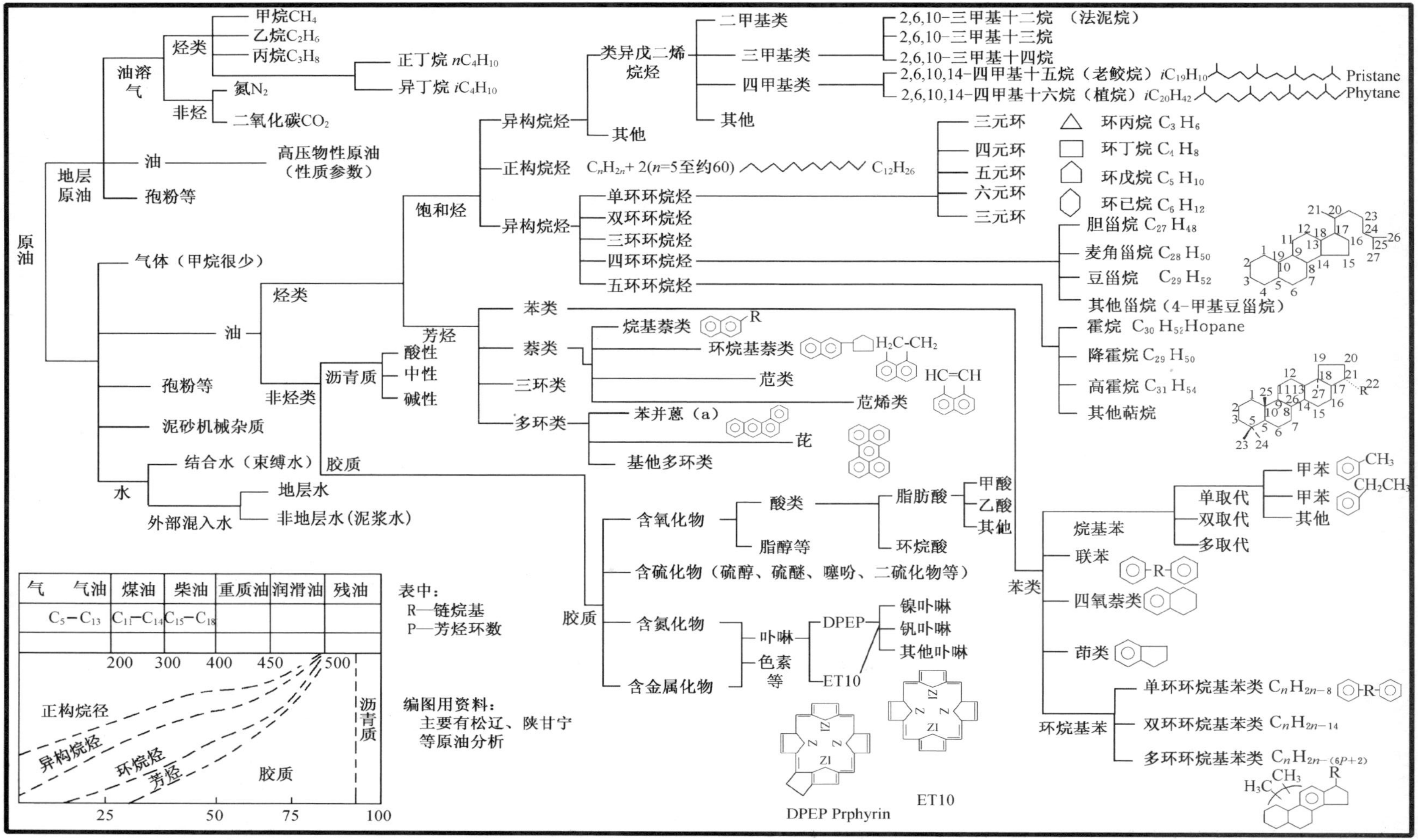

# 附件(Attachment)

## Ⅰ. About the author

### Introduction

He Zhigao, Senior geochemist, born in 1935. He graduated from Nanjing Geological School in 1953 and studied in Northwest China Agriculture University in 1957. His major researches focus on the oil/gas origin in No. 3 Petroleum Prospecting Party during 1960—1973, as head of Lab, in Research Party of Petroleum Geology, MGMR during 1974—1976 as a deputy director and director of the Lab. He served for central Lab of Petroleum Geology, MGMR from 1977, and has been the head of Lab and chief geochemist from 1984. He has written about 30 papers and reports published in different journals (the main papers see the list unclosed).

### Published works of He Zhigao

(end off 1993)

1. Prospecting and Discovery of Oil and Gas in China. Geological Publishing House, 1992. (in Chinese, One of the Contributors)

2. A Model of Oil-generation in a Waterlogged and Closed System. Joural of Southeast Asian Earth Sciences, vol. 5, 1991, P. 465 -469.

3. On the Origin of Petroleum Hydrocarbon Gases. Acta Geologica Sinica, vol. 60, 1986, P. 57 -65.

4. Thermochemical Method of Calculation for Gas Reserve Generated from Coal Measures, in Exploration of Natural Gas. The Petroleum Industry Press, 1986, P. 200 -203. (in Chinese)

5. Composition Characteristics of Natural Gas and the Mechanism of Its Genesis. Experimental Petroleum Geology, vol. 7, No. 4, 1985, P. 276 -282. (in Chinese)

6. Discussion on the Evolution of Hydrocarbons in the Light of the Theory of

Thermodynamics. Exprimental Petroleum Geology, vol. 4, No. 1, 1982, P. 1 - 9. (in Chinese)

7. A Theory for Petroleum Generation Under Integrated Effects of Aqueous Close System. Experimental Petroleum Geology, vol. 10, No. 4, P. 315 - 324. (in Chinese)

8. Basic Geochemical Characteristic of Source Rocks in Songliao and Other Basins of Non-Marine Facies. Experimental Petroleum Geology, vol. 3, No. 3, 1981, P. 197 - 206. (in Chinese)

9. A New Approach for Methodology of Petroleum Geochemistry in Carbonated Rocks. Experimental Petroleum Geology, No. 2, 1983, P. 160 - 164. (in Chinese)

10. Application of Infrared Spectrophotometry in Studying the Evolution of Kerogen. Experimental Petroleum Geology, No. 4, 1981, P. 307 - 316. (in Chinese)

11. On the Origin of Oil-gas from Their Distinction of Property and Composition in West—Taiwan Basin. Taiwan Petroleum Newsletter, No. 2, 1992, P. 1 - 5. (in Chinese)

12. On the Oil-gas Property and Exploratory Direction of Taiwan by the New Theory of Petroleum Origin. Taiwan Petroleum Newsletter, No. 2, 1988, P. 105 - 108. (in Chinese)

13. A New Method of Enhanced Oil Recovery—MFA. Joural of The Petroleum Knowledge, No. 5, 1992, P. 45. (in Chinese)

14. Importance of Studying Crude Oil Upon Itself. Jornal of The Petroleum Knowledge, No. 3, 1991, P. 6. (in Chinese)

15. Appraisal of Carbonate Source Rock Using Gaseous Hydrocarbons Released Acid Treatment. Experimental Petroleum Geology, No. 2, 1983, P. 160 - 164. (in Chinese)

16. On the Reductive Coefficient of Iron. Petroleum Geology Experiment, No. 1, 1965, P. 1 - 7. (in Chinese)

17. Determination of Sulphur of $FeS_2$ in Molybdenite. Northwest Experiment, 1959, P. 75 - 77. (in Chinese)

(After 1993)

1. Mudrock Not Being Oil-Gas Source Rock. (The Paper, 30th International Geology Congress)

2. Evidences of Petroleum Hydrocarbon Gases Directly by Microbial Fermentation. (The Paper, 30 th International Geology Congress)

3. Table of Crude Oil Component. The Petroleum Knowledge, No. 4. 1996, P. 24 – 25. (in Chinese)

4. On Evil of the Theory of Oil-generation with the Thermal Degradation of Kerogen. Marine Origin Petroleum Geology, No. 2, 1999, P. 50 – 56. (in Chinese)

5. Thought and Comment on Some Problems in the Oil Origin Theory. Marine Origin Petroleum Geology, No. 3 – 4, 2000, P. 118 – 122. (in Chinese)

6. The Non-marine Petroleum and the Theory of Petroleum Origin. The Petroleum Knowledge, No. 3, 2003, P. 32 – 33. (in Chinese)

7. "The Chinese Ghawar Embryo" and Irrefutable Section of the organic origin of petroleum; "The Olefiant Marstone Rock" named representations. The Petroleum Knowledge, No. 3, 2011, P. 4 – 7. (in Chinese)

8. The Idea to Form a Next Oil Trap in the Old Oil Field for Getting Unrecovered Oil. The Petroleum Knowledge, No. 4, 2007, P. 6 – 7. (in Chinese)

## Non-Publications (非公开出版物)

1. Experimental Basis of Petroleum Geology (Infrared Spectroscopy; Porphyrin; Crude oil etc) Books, about 150 (Chinese word) thousands words, in 1979.

2. The Quality Management Manual of Ministry of Petroleum Geology Central Laboratory. Data, about 50 thousands words (Chinese word), in 1993

3. Textbooks of Petroleum Geochemistry. Data, about 200 (Chinese word) thousands words.

4. Research Reports (Departmeant of Geology "Petroleum Geology Research Battle, in Hunan"), about 50 (Chinese word) thousands words, in 1981.

**Russian translations, data, about 100 (Chinese) thousands words.**

## Ⅱ. 何志高十项创新科研成果简录

## Ten brief record of innovation scientific and technology achievements by He Zhigao

一、1959 年,辉钼矿中黄铁矿形态硫相分析方法创新,为我国重要钼矿——金堆城钼矿的选矿做出贡献,方法载于《西北实验》(1960 版)。

二、1964 年,创立沉积地球化学相指示标志——铁的还原系数 $x_{Fe}$,同时批判了苏联权威地球化学家齐诺维也夫的 $K_{Fe}$,全文发表在《石油地质实验》1965 年第 1 期第 1 篇,还有编者按及第 2 篇笔者(陈丕济、陈焕疆、张义纲 3 位教授)的前言评价,可见学术价值和水平是当时年代国内最高的。

三、1976—1980 年,红外光谱上的创新有两个方面,一是芳烃结构分布指数中光谱谱带的归属,二是干酪根红外光谱的芳环位移律的发现。(有两篇对应论文。)

四、创立一种新的天然气组成特点与成因机制的热力学-数学分析方法与理论,具有独特的新颖性、开创性,全文载于《石油实验地质》1985 年第 4 期第 276—281 页。

五、创立一种煤成气量的热力学计算方法及生气率 $Q$a 值($Q$a = 187 m$^3$/t 煤)。该数值与十多年后我国学者赴美考察所得到的"煤化过程中每吨成熟煤大约可生成 198 m$^3$ 天然气"之值相当接近,也检验了该计算方法的准确性和所创立的计算理论的科学价值。发表在石油地质进展丛书 3《天然气勘探》(第 200—203 页,1986 年 9 月出版,北京)。

六、创立一种石油天然气的成因理论,全文中、英文两种版本刊载于《地质学报》1986 年第 4 期(《论石油烃类气体成因》第 365—373 页; *On The Origin of petroleum Hydrocarbon Gases* 第 57—65 页)。

七、创立一种石油成因新理论——湿封闭体系微生物和碳酸盐主导成油论,第一篇发表在《石油实验地质》1988 年第 4 期;在第三届国际石油勘探及地球化学大会报告(地质矿产部唯一的一篇);由 Journal of Southeast Asian Earth Sciences 发表题为"A model of oil-generation in a waterlogged and closed system" vol. Nos 1 – 4, P. 465 – 469, 1991.

该理论受到国外学者关注,英国伦敦皇家学会前任会长 R. Robinson、美国 TEXACO E&P TECHNOLOGY DIVISION Core Robison 来函索取论文。1984

年,新疆沙参二井喷油,当年,其油源报告在该理论指导下,国内外最早第一个由担任主任工程师的何志高明确提出属海相古生界油源,20 多年的历史证明该结论是正确的。

八、世界上首次系统、全面、深刻、有力地批判统治学术界近半个世纪的《干酪根热降解成油论》全文发表在《海相油气地质》1999 年第 4 期。

九、"煤不成油"、"泥质岩不是生油岩"等重大石油地质领域的重要创新理论,都将对整个石油地质科学理论与实践具有重要意义和震撼性。发表相应论文 5 篇。

十、发明专利,一种提高原油采收率的新方法——马弗阿法。1992,专利号第 13321 号。

# Ⅲ. Letter from AAPG

## American Association of Petroleum Geologists

An International Geological Organization

AAPG Bulletin
*Fax*: *(918) 584 - 2453*
December 20, 1994
He Zhigao
P. O. Box 916
Wuxi, Jiangsu
People's Republic of China

HE AND SHEN
AAPG Ms. No. 218

Dear Dr. He:

I have received your manuscript, "Mudrock Not Being Oil-Gas Source Rock."

Manuscripts submitted to the *AAPG Bulletin* are reviewed by the AAPG elected editor, Dr. Kevin T. Biddle (Exxon Exploration Company, Houston, Texas), and three reviewers consisting of *Bulletin* associate editors and/or invited reviewers. Dr. Biddle will make the final decision of acceptance or rejection of the paper.

The *AAPG Bulletin* does not accept papers that have been published or that are being considered elsewhere. Authors are responsible for ensuring that the data to be published have been released, and for securing permission to use previously published material (table, figure, etc.) in the paper.

As you know, we print full-color photographs, reproduced from slides, on the *AAPG Bulletin* cover. If you have any slides that you think might make an attractive and interesting *Bulletin* cover, we would be more than happy to consider them. A cover photo pertaining to your manuscript would help increase the interest in your article. Do not send slides that would duplicate figures in your paper. If a slide is not available, we can use a high-quality photograph. Please note that your slide or photo will not be returned.

Please provide an E-mail address for use in corresponding on the status of your manuscript.

Thank you very much for submitting your paper to the American Association of Petroleum Geologists for consideration.

Sincerely,

Kenneth M. Wolgemuth
Publications Manager

KMW/pw

# Ⅳ. 关于石油成因新理论应用情况与前景的报告

**The report that is the petroleum origin new theory by the author's work unit**

## 石油局科技处:

我室何志高同志的石油成因新理论研究,自1986年首篇论文在《地质学报》(中英文本)发表以来,至今十年。其主篇,1989年第二届亚非石油勘探会议被选为大会发言论文(地矿部唯一的一篇),1991年*Joural of Seas*刊出全文。1994年,大庆油田微生物采油室的实验证明,生化反应直接形成大量的石油烃类气(我们提供的微生物),有力地验证了石油成因新理论。根据该实验资料撰写的论文摘要已被第30届国际地质大会编入文集。此外,还有论文数篇,扩大和深化了石油成因新理论。

### 一、主要应用情况

1. 1987年,中原油田临清坳陷生油研究,应用新理论提出"雏形油气圈闭"的概念与实例,总报告已获中原油田一等奖,报告已成书。

2. 1985年,新疆沙参二井喷油。当年,其油源报告在该理论指导下,国内外最早明确提出属海相古生代油源。十多年的历史证明,该结论是正确的。

3. 台湾台西盆地油气组成与性质差异成因分析,《台湾石油通讯》以第一篇文章刊载于1992年第2期。

4. 国家攻关课题75—54—01—01中一个3级专题,二氧化碳成因与油气关系,应用新理论所提出的结论已被张义纲主编入《天然气生成聚集与保存》一书。

5. 苏北F第三系油源专题报告,明确指出戴南组原油不是来自下部阜宁组,而是自生自储。打破传统观念,受油田赞扬。

6. 我国南方中古生代海相碳酸盐岩的研究指出,石油的富集带是相变过渡带。1987年中国-澳大利亚石油代表团学术交流期间,全国第三届地化会议论文集上发表。

## 二、国内外评价与反馈

1. 英国伦敦皇家学会前任会长 R. Robinson 来函索取论文，当收到后，来函表示一定认真研究。

2. 美国 TEXACO E&P TECHNOLOGY DIVISION CORE Robison 来函索取论文。

3. 澳大利亚石油代表团来华参加中澳学术交流期间，澳方学者认为，报告很好，很有说服力，回去要深入研究。

4. 马来西亚《地球科学杂志》主编 K. Boo 来函说，论文内容很好，对理论很有兴趣。

5. 美国 *AAPC* 杂志编辑对泥质岩不是生油岩的论点很赞赏。

6. *Mine* 杂志刊登详细摘要。

7. 《石油知识》(1995.4)刊登题为《发展石油成因新理论刻不容缓》一文。

8. 石油知识组织评选，将《湿封闭体系综合效应成油论》一文评为优秀作品，并汇编在《石油知识文萃》一书中。

9. 受中南石油地质局、吉林油田之邀请，讲授了石油成因新理论，获一致好评。

10. 国内个人评价中，致函评价很高者有地矿部北京地质所总工谢秋元、石油大学副教授劳永新等，化探中心高级工程师张金来认为新理论对油气化探指标，尤其是 ΔC 的机制解释很有意义。

## 三、前景

新理论是一崭新的理论体系，具备热力学理论基础，可望最终解决自然科学中这一复杂难题，在国际科学技术，尤其是在石油地质科学上肯定具有重要意义。当前，各地区(新老油气区)都可应用；大庆深部的油气普查勘探、我国西部的油气勘探以至南方中古生代海相碳酸盐岩均会起到重要的作用。

但是，由于新理论是对以往各种传统理论不同程度的否定，阻力很大，难以被人们接受。

地矿部石油地质中心实验室

一九九六年六月[illegible]十日

# V. 报纸上的文章
# Newspaper article

## 石油地质中心实验室何志高提出石油成因新理论
### ——湿封闭体系综合效应成油论

**本报讯** 最近,《石油知识》杂志以《发展石油成因新理论刻不容缓》为题,发表了地矿部石油地质中心实验室高级工程师何志高建议应用他所创立的一种石油成因新理论——湿封闭体系综合效应成油论,阐述了"我国石油储量、产量欲大增,石油成因理论必须更新"这一新观点,再次引起了同行专家们对他的这一新理论的关注。

作为一种特殊的流动性的矿产资源——石油,是由以很多系列化合物为特征的烃类和非烃类的复杂混合物组成的。正是基于这种流动性、易变性、组成的复杂性以及赋存规律的特殊性,使得石油成因研究经历了无机与有机、早期与晚期、"干"与"湿"等理论上的为时一个多世纪的长期争鸣。长期从事石油地质与实验测试研究工作的高级工程师何志高,从20世纪70年代初开始,深入油田实践,系统地研究了油田宏观、微观的大量实际资料,掌握了各种近代分析测试技术所提供的比以往更加准确、更加广泛的定性、定量组成分析的信息,引进微生物发酵研究领域的成果,应用热力学理论,发现了沉积有机物质和重新分配作用与脂肪酸的脱羧规律,在理论与实践紧密相结合的基础上,历经20余年潜心研究,最终科学地提出了石油生成的时间、地点、条件与机制。

何志高创立的这一新理论认为,石油形成于湿的(饱含水的)封闭体系之中。该体系的特点是多孔隙、饱和水。例如,砂岩(砂质沉积物)接受了下伏泥质沉积物由微生物改造和物理(轻重分异)的作用再分配而富集于表层的细分散状、胶体状和可溶性的有机物质,再加上砂质沉积物本身随同沉积的有机物质,使得该饱和水的、多孔隙的体系聚集着丰富的、生油效率十分高的有机物质。当泥质沉积物再覆盖砂质沉积物作为盖层之时,这种体系就成了单一的湿封闭体系。在这种湿封闭体系中,首先形成各种脂肪酸类、类石油胶质、类石油沥青以及甲烷等,然后进入成油聚集阶段。在成油聚集阶段,在饱含水的封闭体系中,继形成脂肪酸等的生化反应之后,开始了脂肪酸的脱羧生化反应,形成各种气态烃类以及各种高分子烃类。这些烃类,加上胶质与沥青质,构成了早期生成的石油。随着埋深作用增强及气态烃、可汽化的烃类等气体的生成，体系中的压力随之增加，温度增高。由于压力和温度的作用，分散的石油聚集成体积不等、含油级别不同的含油体。温度的另一种作用是使类石油胶质、类石油沥青质降解形成一部分烃类与非烃类。这部分烃类与生物作用直接形成的烃类构成了湿封闭体系中全部石油烃类之总和。这些全部烃类与胶质、沥青质在压力的作用下，气态烃进一步液化，石油各组分分子之间内聚力增强，并受岩性等条件的影响，最终形成规模不等的油藏。在成油过程中，存在一种极为有趣的作用，即脱羧等生化反应生成二氧化碳，在有水的条件下发生反应，使得碳酸盐胶结物溶解，从而为同时生成的烃类创造了储集空间。

何志高创立的这一新理论还认为，在碳酸盐岩沉积区，石油形成于一种“窄而厚的相变过渡带”，即在横向上，碳酸盐沉积物与碎屑沉积物的相变过渡带是富集有机物质的、孔隙发育的、含水的地质体。根据他的研究表明，这种过渡带相当之窄，通常仅在数千米的限度之内。从这一理论出发，何志高强调我国约200万平方千米的南方中古生代碳酸盐岩地区的石油普查勘探必须把基点放在这种相变过渡带，只有这样，才能有重大突破。对大庆深部碳酸盐岩地层、塔里木海相碳酸盐岩地层以及塔里木全盆地的油气开发等亦如此。

将石油的生成、运移与聚集有机地融为一体的这一石油成因新理论，曾在第二届亚非石油勘探与地球化学会议和全国第三届有机地化会议上宣读，并在《地质学报》、《东南亚地球科学》等国内外杂志上发表，受到了国际国内的关注。这一新理论曾被用来对台湾台西盆地的油气田进行了有效的研究，在临清坳陷，提出了“雏形油气圈闭”的实例，不仅解释了包括“不成熟”原油、稠油的形成机制以及一系列石油地质问题，而且以新理论的基本原理，创立了一种可提高原油采收率的新方法——马弗阿法，该方法获得国家发明专利证书。到目前为止，何志高应用这一新理论，先后完成九份专题研究报告和文章，分别在有关全国学术会议和《石油实验地质》、《台湾石油通讯》杂志进行了交流。

我国原油年产量已经达到1.4亿吨，这是了不起的业绩。数十年的全国油气普查勘探开发的实践和石油成因理论的研究表明，油气储量、产量欲大幅度地增加，石油成因理论必须更新，热心于石油成因新理论研究的何志高深信，石油成因新理论一旦被石油地质学家所掌握，必能发展石油地质科学，为全球尽早地开发出更多的油气资源。

无锡石油地质中心实验室
江其勤

# 跋:简评“石油无机成因论”

# Postscript: Comment on “Petroleum inorganic origin theory”

近年来,中国的“石油无机成因论”兴起之势与20世纪五六十年代苏联的情况颇为相似,笔者还有记忆。笔者认为,石油无机成因论者的精神是值得本人学习和称赞的。但是,该论者可能不太了解石油组成成分的复杂性和特殊的物理化学性质以及其地球化学行为,尤其缺乏原油的对比研究。笔者曾经发表过一篇短文,题为《对原油本身研究的重要性》,刊于《石油知识》1991年第3期,该文后来又被收入石油工业出版社出版的《石油知识文萃》一书中。

笔者推荐以下两种对比方法:一是酸解含油碳酸盐(油泥灰岩)对比法。该法相当于传统的沥青A和沥青C的对比。二是海相原油与陆相原油(包括膏盐沉积盆地原油)组成分析测试对比法。

岩石对比样品:首选中东地区,例如加瓦尔油田的“生油岩”和“储油岩”各1份,约1 kg,按历来的岩石分析方法进行样品加工。

原油样品:(1) 中东海相原油;(2) 大庆陆相原油;(3) 中国江汉油田原油。

除了上述3种原油外,还可以选取“费托法”人工合成的油一并分析测试。

分析测试项目:

(1) 元素分析:含硫量,钒、镍含量。

(2) 族组成分析:主要是芳香烃含量。

(3) 常规地化指标分析:饱和烃色谱分析及相应OEP、CPI等参数;芳烃结构分布指数。

(4) 生物标记化合物的分析测试。

(5) 稳定碳同位素分析测试。

结果结论评估:

岩石中的石油液(流)体可以通过孔隙(包括喉道)中的石油与结晶体(分子)之间的包裹石油,两者具有高度相关性。这是无机成因根本无法解答的致命问题。

陆相原油和海相原油组成成分与结构的巨大差异以及原油的组成结构特征特性与沉积环境密切相关,这也是无机成因论的另一个无法自圆其说的问题。

笔者最后还想表达的是:历史上的有机成因理论最大的失误是把“富含有机物质的泥岩是生油岩”作为一切理论和实践的基础,先入为主。事实上,泥岩不是生油岩。

新的时代开始了,湿封闭体系微生物和碳酸盐主导成油理论必将屹立于世界科学殿堂。

# 祝贺与致谢
# Congratulations and the author's acknowlegements

拜悉姐夫大作面世,恭喜!

早在20世纪70年代初,我们就相知,时任地矿部三普实验室主任兼技术负责的小姐夫非常敬业,刻苦读书,其钻研精神可敬可佩,40多年后终成正果。出版此书,也将惠及我们家族,利及后人发扬光大之。

内弟　吴宏昌
弟媳　费惠君
于苏州
2013年10月

笔者特别致谢(The author's acknowledgements):

惠君是苏州江南嘉捷电梯股份有限公司副总裁,宏昌是苏州德意机电设备有限公司总经理,他们都是事业集大成者,赞助姐夫出版此书,亲情厚眷,特致感谢。

# 一碑
# （A monument）

何志高同志一生奉献给了石油地质事业，在科研道路上兢兢业业耕耘了五十余载……

张伯荣　（原长庆油田党委书记）
原文刊登在《石油知识》2007 年第 4 期